Fetch Felix

The Fight Against The Ulster Bombers
1976 – 1977

★

LIEUTENANT-COLONEL
DERRICK PATRICK
OBE

HAMISH HAMILTON
LONDON

First published in Great Britain 1981
by Hamish Hamilton Ltd
Garden House 57–59 Long Acre London WC2E 9JZ

Thanks are due to William Collins, Sons and Company for permission to
quote from Edmund Blunden's poem *Escape* in *Undertones of War*, London
1965 and to John Murray for permission to quote from *A Place Apart* by
Dervla Murphy, London 1978

Patrick, Derrick
  Fetch Felix
  1. Bombings – Northern Ireland
  2. Bomb reconnaissance – Northern Ireland
  I. Title
  358'.2   HV6640
ISBN 0–241–10371–1

Photoset, printed and bound in Great Britain
by Redwood Burn Limited, Trowbridge & Esher

Contents

A map of Northern Ireland appears on page viii and a section of illustrations follows page 88.

Foreword

The idea of writing this book was given to me one evening in the Officers' Mess Bar at Headquarters Northern Ireland. It was about the time of the Queen's visit as part of the Silver Jubilee celebrations. The suggestion was made by a group of officers and civilians who were, I suppose, only onlookers on the bomb scene, but had nevertheless played a major part in helping us to deter the bombers – the Operations Staff, the Watchkeepers, the Royal Air Force and Army Air Corps, the civilian scientists, all the supporting services and the Public Relations staff. It was their enthusiasm for the events to be recorded that has led to this book being written.

A special word must be said for Bill Moore. At the time that I was Chief Ammunition Technical Officer, Northern Ireland, he was the Public Relations Officer for the Ulster Defence Regiment. He is also a military historian of some repute. He has collaborated fully with me and spent many hours editing the story and putting it into readable form. Without him and his professional expertise I doubt whether *Fetch Felix* would ever have seen the light of day.

The incidents that occurred during my fourteen months in Northern Ireland were many and varied, and I have selected only a handful to put down on paper. However, and rather inevitably, even the bomb disposal personnel and intelligence officers who played a direct part in the events I have chronicled and are still serving cannot be named. All have had to remain unsung. But thirteen members of the EOD sections were decorated during my tour, and they were all very brave and selfless men. To borrow that ominous phrase, they know who they are – and I would wish to pay tribute to all of them for their dedication, devotion to duty and high standards of professional skill which, at the end of the day, contributed to the very high success rate we achieved against the terrorist bomber.

D.P., 1980

Portrush

Coleraine

ANTRIM

LONDONDERRY

Garvagh

LONDONDERRY

Dungiven ⫛ Glenshane
Pass

Ballymena

Foyle

Strabane

Bellaghy

Glenelly

Magherafelt

Moneymore

TYRONE

Aldergrove ⌂
Airport

Bann

Ligoniel

Lough
Neagh

BELFAST

Bangor

Omagh

Dunmurry

Dungannon

Lurgan

LISBURN
'HQ Northern Ireland &
base of 321 EOD Unit'

Blackwater

Portadown

Hillsborough

Lagan

FERMANAGH

Enniskillen

Armagh

ARMAGH

Bann

DOWN

Kinawley

Newtownbutler

Newry

Crossmaglen

Jonesborough

Warren Point

Carlingford Lough

0   5   15   30 miles

5   20   40 km

CHAPTER ONE

'A Merry Christmas to You Too'

According to Warrant Officer A—, who was watching from behind cover, my legs were churning through the motions of running before I hit the ground. Until a few moments earlier everything had been proceeding in slow time. He'd watched me reach carefully into the chimney and fumble around. He'd seen me ease out the black object, lay it gingerly across the lip of the stack and peer at it like a plumber who has removed something unpleasant from a drain. At that point the scene froze. I stood silhouetted against the grey February sky, A— remained motionless beside our vehicle, and the bomb team, sensing a change in mood, stopped brewing up. The next move was up to me.

I came off the roof of that bungalow in a flash. I didn't attempt to make for the ladder or worry about what sort of impact I might make on the ground. I plumped down on my behind and slithered feet first down the tiles, shot over the gutter, crashed to the ground and bolted.

The faces of the bomb team showed polite curiosity as I skidded into their midst. Bombs they were used to, but lieutenant-colonels who moved like greased lightning. . . .

'The bloody clock has started to tick,' I said. 'It's bloody-well ticking now!'

The silence was broken by the warrant officer, a short, smart man in his 30s.

'Well, at least we know why we're called Felix. A cat couldn't have come off that roof any faster.'

This sally gave the bomb team the excuse they had been waiting for and they laughed outright. Clearly it was a highly amusing incident.

'See 'is face as 'e went over the gutters?' I heard one of them say, as they went back to their tea-making. 'Best thing since sliced bread.'

I looked for inspiration at the red and white Tom and Jerry-style cat insignia on the armoured Ford Transit vans. None was forthcoming. Felix! Bugger Felix! There was a bomb ticking away on top of the chimney and it had to be dealt with. More precisely, *I* was obliged to 'neutralise' it.

A— stared at the menacing black object. He was a patient man, fully aware of the realities concealed behind the word 'neutralise'.

'Anyway, sir, at least we've got the beggar where we want him now.'

It was about time.

For me, the search for that particular bomb began one night in February 1977 when Headquarters 3 Brigade, at Porta-down, reported 'a problem' near Omagh, in County Tyrone. H Jones, the brigade major, told me that they had received a tip that a bomb had been planted in a bungalow occupied by an RUC reservist and his wife. Would I contact the Special Branch down there? Apparently there were complications.

'What sort of complications?'

H Jones sounded cheerful. 'I think it would be better if you got the whole story when you got there, Colonel.'

'Hmm. OK. I'll go in the morning.'

'Goodnight, Colonel'.

Goodnight to you too, H Jones! The mystery of the so-called complications would have to be solved in the morning. I sent word to Lance-Corporal Coupe, my driver, that we would be leaving soon after 9 am.

It proved to be a cold morning and the Chunkies* on the vehicle check-point which controls all traffic in and out of HQ Northern Ireland were stamping about to keep warm. Coupe, a Jeeves-like individual from the Royal Corps of Transport, nosed our Hillman Hunter gently over the automatic ramps which at that time were the subject of considerable specula-tion. Large metal wedges set in the road, they could be raised if required to halt traffic at the push of a button. But, from time to time, they seemed to have a mind of their own, springing up under moving vehicles and leaving hapless drivers bumped and bruised while their vehicles drooped alarmingly at both ends. Victims always vigorously denied any culpability, such

* Chunky – a nickname for men of the Royal Pioneer Corps.

2

as ignoring traffic signals, but received scant sympathy. So Coupe, who had no wish to join the Ramp Riders' Club, negotiated the exit carefully and drove us smoothly down the hill, past the neat Victorian villas and school, to the centre of Lisburn where we merged with the bustling traffic.

Outwardly there was nothing to indicate we had anything to do with the Security Forces. Our Hunter had civilian number plates, differing from normal cars only by reason of the two-way radio which was not visible to outsiders. There were three of us in it, as I had asked the major commanding 321 Explosive Ordnance Disposal (EOD) unit at Lisburn, to come with me. We wore civilian clothes.

At Omagh, a busy little town with a sizeable bus station which had attracted the attention of terrorists on occasions, we made for the headquarters of the 9th/12th Lancers, then serving an 18-month tour in Northern Ireland. The local bomb team was located with the 9th/12th and we picked up W02 A— who was in command. He knew the Special Branch sergeant from previous dealings and was able to fill in some of the background. The team followed in their bomb van.

The RUC Special Branch sergeant was waiting for us at the police station, young, dark and confident. He pointed to a spot on the map.

'Our information is that there is a bomb in there, Colonel.'

A solitary building was marked by a red circle on the talc.

'We haven't seen it, but I'm sure it's there, all right. Our source is pretty good – usually dead on.'

Most sound opinions in Northern Ireland are 'dead on'.

'The couple who live there are both in the police reserve.'

The bungalow was a few miles out of town with the nearest dwelling some distance away. Inwardly, I took my hat off to two people who were brave enough to live in isolation in a troubled area.

'What about these complications I've been told about?'

The Special Branch man grimaced.

'Well, it's the old business. We'd like to clear the place with the minimum amount of fuss. It's our informant. We've got problems and we'd like to do this as quietly as possible.'

The light was beginning to dawn.

The Special Branch man continued. 'We asked the Royal Engineers if they would search the bungalow for us and see if there was a bomb inside, but they wanted the Army to cordon the place. That would cause too much fuss and take days. As I say, it'd suit our book if you could do it nice and quiet like.'

I could see the sergeant's point. A normal operation takes time. There has to be an inner cordon and an outer cordon with as much as a company of troops employed – the men on the ground have to be relieved, don't forget. Aerial pictures are taken before and after the Royal Engineers have cleared the paths and the immediate environs of a target before the EOD (Explosive Ordnance Disposal) team will move in.

But the Sappers had a point too. The whole thing might be a 'come-on' – an incident organised to lure members of the Security Forces into the area where they could be fired on by snipers or blown up by hidden explosives. You couldn't blame the Royal Engineers for wanting to stick to the book.

'You're sure the couple are the target, sergeant?'

He nodded.

'OK, let's talk to them.'

The man was in another part of the station but his wife, a woman in her twenties, had to be brought from the shop where she worked. Both were very nervous and unable to help. They hadn't seen anyone about the bungalow, which they rented, and didn't have a clue where the supposed bomb might be.The wife, however, was as certain as the Special Branch man that there was one there.

'It's the dog, sir. It's never been restless before but it sud-denly got so it couldn't wait to get out of the place. It doesn't go for the place at all. There is something in there now that wasn't there before.Something wicked.'

She wore a short duffel coat on top of her overall and rubbed her hands against the cold.

'That dog knows something, I'm certain sure, sir.'

There seemed to be only one way of finding out what was bothering the dog. Taking the wife with us, we drove over to the bungalow – in just the one car to avoid suspicion. The warrant officer went back to barracks with the bomb team's van. Anyone watching would think he had been on a routine call. We hoped no one would connect him with us.

The bungalow lay up a narrow winding road and was set on a slight eminence. There was nothing particularly impressive about the place but for me it wore the challenging, baleful air that suspect buildings assume ... dumb insolence in the shape of bricks and mortar. We spent some time looking round the area before we decided to go in. When we did it was by the back door. Once that was open we could go through the house by pulling open the doors from a distance instead of pushing them close to.

There was no problem with the door, but the kitchen felt chill and unfriendly. It seemed to resent our methodical intrusion into its contents.

I had decided that we would clear the house together, Geoff and I: partly because it might be too much of a strain for one person, partly because we didn't want to be exposed afterwards to any charges of causing unnecessary damage or stealing things. Stranger things have happened and it was as well to have each other as witnesses.

The presence of the wife who sat in the car with Coupe while we worked was essential. Even though the place was rented the woman knew just what was in it. As we searched we went to and fro asking her questions. What sort of detergent did she use? Where did she keep her tea? Were there any firelighters in the kitchen? Where a husband would have had to think she answered without hesitation. Gradually we established that nothing alien had been introduced into the kitchen and that nothing was out of place.

We began to go through the other rooms. Happily the door handles were of the lever type and we were able to pull them down with a hook and line before pulling open the door with tape.

The fact that the couple were not the tidiest folk in the world did not help matters. The bed had not been made and we had to use a hook and line to pull back the blankets. Once Geoff turned to me with raised eyebrows and held up a pair of panties he had discovered while rummaging under the settee.

In sequence we cleared the kitchen, sitting room, bathroom, and toilet before opening the door to the spare room.

'Christ, this place is in a state,' said Geoff, a mild-mannered, medium-sized major with a clipped dark moustache who was always a model of neatness himself.

'Where do we look now?'

A pile of junk filled the place. There were rolls of carpet, boxes, old household gadgets, lampshades, the lot. It took us longer to go through that room than any of the others, unrolling the carpet, systematically opening the boxes.

The cold was striking deep into our bones by now, and we were beginning to have our doubts. By the time we had finished we had searched every corner of the bungalow and found not a scrap of evidence, not a whiff of anything resembling explosives.

'It's all right as far as I'm concerned, Mrs —,' I said when we went back to the car. 'You can go back in. There's nothing there.'

From her face it was clear she only half believed us, but she began to apologise for taking up our time, and, woman-like, for the state of the house – it had been very plain that they had left the bungalow in one devil of a hurry.

We told her it was all in a day's work and took her back to Omagh. The husband looked relieved but the Special Branch man shook his head.

'Our informant hasn't been one to put us on the wrong track before. Still, there has to be a first time.'

Clutching a mug of coffee in my numb fingers, I agreed. Later we drove over to tell our tale to the ops officer of the 9th/12th Lancers, then in residence at St Lucia Barracks, Omagh, a splendid grey old Victorian stronghold once the home of the Royal Inniskilling Fusiliers. Mementoes of the Boer War hinted at orthodox soldiering in better days in warmer climes. I phoned H Jones at 3 Brigade and told him it was a false alarm.

Two days later the phone rang in the Portakabin I used as an office. It was the Special Branch man at Omagh. He sounded very cheerful and very Northern Irish.

'Colonel Pa-atrick? It's there!'

So he'd found . . . what? Where?

'There's something down the chimney that shouldn't be there,' he replied triumphantly. 'I've been up on the roof and had a look. I shone my light on it. I'm certain it's the baarm, all right. Dead on, as we were told.'

What was not dead on was the next suggestion that I should send someone down to extract the object by night. The RUC were still keen to remove it with us as little fuss as possible. I said that while I appreciated their concern for their informant,

6

I had to take a number of things into consideration. If I were to operate on the roof at night it would have to be lit and I would make an excellent target without anyone being able to spot terrorists likely to fire at me. Secondly, the amount of lighting I would need would attract probably the very interest they wished to avoid.

There was a vague hint of disappointment in the sergeant's voice as he conceded my point and accepted that I would be with him on the morrow.

Protocol then needed to be observed. As I was going to be operating in 3 Brigade's area I needed their consent to all my actions. I called up the brigade major on a scrambler net – as we could not be sure the normal military line was secure – and explained my plan to him. Agreement followed with almost disheartening rapdity.

The next day Coupe and I set off alone for Omagh where we were to meet A— and his bomb team plus an escort of the 9th/12th Lancers. Because of the complications and the need to do the job quickly, I had decided against calling for extended cover while the task was carried out. From the evidence available I was convinced there was no need to mount a major conventional operation using helicopter surveillance and a big stake-out with troops.

The RUC had been back to the bungalow and the family, too, dog and all, so it did not appear to be a come-on for the Security Forces. It might be an attempt to kill me, of course. That was always on the cards, though I felt it unlikely in this case. No, this was a straightforward, nasty attempt to murder a man and his wife with a bomb. They were undoubtedly the target of the terrorists. In my opinion the area around the bungalow was safe, but if things did not turn out quite as expected at least I had a full bomb team and a certain amount of firepower at my disposal.

On the way out to the bungalow I travelled with the warrant officer in the lead vehicle. We didn't mention the bomb during the ride, though the weather came in for a good deal of discussion.

According to the enthusiastic Special Branch sergeant, the chimney the bomber had chosen served the fireplace in the spare room. I had looked up during our earlier search but the flue had been stuffed with newspapers. The wife had confirmed they had blocked it to stop draughts.

After we had parked the vehicles, I studied both of the bungalow's chimneys carefully. They stood about three feet higher than the ridge of the roof and were made of brick without any pots. Obviously any approach would have to be via the roof. My first decision was made for me . . . I would not be able to wear a bomb suit. Bomb suits with helmet, breast plate, and padding, are designed to protect operators from some of the effects of explosions – blast, burns, debris. They are strong, but heavy. I could see no way in which I could climb a ladder and then crawl about the tiles in one. I would have to go up as I was, in camouflaged combat jacket, green combat trousers, puttees and 'boots DMS', the direct moulded sole footwear which the Army wears today in place of the good old ammunition boot. The beauty of the DMS boot is that it is without studs and has a good grip on a slippery surface. You have no fear of striking unwanted sparks.

On my first approach, up a ladder placed at the rear of the bungalow – the one which had been used by the enthusiastic sergeant – I took only a torch. With this I looked first down the chimney we believed to be clear, to make sure that it was and to get some idea of the way in which the builder had worked. Then, straddling the ridge, I worked myself along to the other.

Whereas the first had smelled faintly and not unpleasantly of peat smoke, it was obvious that the second had not been used for some time. The bird droppings around the rim showed that. It was another of those grey and dreary days and what daylight there was did not help people who wished to look down non-functional chimneys. I switched on my flash-light and directed the beam around the interior walls, finally bringing it to rest on something which seemed to have a dull shine. It took me some time to work out that it was a bundle of farm baling wire. Below it there was something else, some sort of parcel with what appeared to be a string handle attached. Having had a good look I climbed down the roof to the ladder and went back to the bomb team.

'It's a funny one,' I said. 'Whoever did the job lowered the device on a length of baling wire until it got stuck in the bend of the chimney. Then he chucked the wire down after it and left.'

The question now arose: How do you get a bomb out of a chimney without breaking the golden rule – that you never,

never pull a bomb towards you? That way lies oblivion.

Clearly, I had no choice but to work on the roof when I attempted to neutralise the device. I could, of course, blow it up – and the chimney, the bunalow and the clues it contained with it. But I wanted the bomb as evidence. So I had to go fishing.

Every bomb team carries an assortment of standard kit, and from it we selected two fishing hooks, the sort you can buy at any tackle shop. These I attached to a length of line which in turn was attached to a fitting known as an Allen hook on the end of a much thicker piece of cord. Thus equipped and under A—'s keen gaze, I returned to the roof. As usual, most of the other members of the team were taking no part in the proceedings at this time and they started to make a cuppa.

My aim was to use chimney No 1 as a pivot for the line. I looped the Allen hook and line round it and then edged along the ridge again to the stack containing the device. This was about four feet below the lip of the chimney, well beyond arm reach, but by persevering with the fishing line I managed to hook the string handle. I straightened the line and signalled to four members of the team to take the strain on the cord dangling from the first chimney. When I was satisfied that it was just the right tension, I came off the roof, we retired to a safe distance where the team was behind cover and I gave the order to pull. All that happened was that the cord broke.

I said a few sharp words about the quality of equipment and returned to the roof. Once again I hooked the string handle and rejoined the four men holding the Allen cord.

'Heave!'

They heaved and this time there was an appreciable movement before the object stuck and the cord went taut. After a moment or two I climbed the ladder yet again and worked my way across the tiles. The flashlight revealed that the device had definitely moved and the problem was being caused by the bend in the chimney. The baling wire was no longer in the way, however, and I felt confident that one good yank would do the trick. Once back with the team I told them that this was it.

The yank did not do the trick. Certainly the bomb moved quite a way – then it stuck. I looked up at the impassive faces of the line-heavers and set off yet again. What all the jolting had done to the device no one could say, but as I scrambled over the tiles I persuaded myself it was not the most sensitive of infernal machines.

My nose was about a foot from the thing when I poked my head over the chimney's edge and stared down. Whatever it was it was black with a dull sheen and had jammed against a ledge running round inside the chimney which was narrower at the top no doubt to encourage the fire to draw well. A white clock was plainly visible beside the baling wire and the string handle.

Handling a live device is something one does as a last resort. There are mechancial means of neutralising a bomb – firing a shotgun cartridge into it for example, to disrupt the mechanism. In this case I considered I had no choice. My knees gripping the stack to balance myself, I reached down until I could get my hands on the string handle. Shouting to the line-heavers to ease the strain a fraction I was able to take the weight of the object. It was heavy.

'Just ease it a fraction more!'

Tilting it clear of the obstructive ledge I brought my prize into the light of day. There is no such thing as a beautiful bomb and this was as ugly as any I have seen, a length of black plastic sewer pipe about 20 inches long and six inches in diameter. I laid it gently so that it was balanced across the mouth of the chimney, which was rectangular rather than square, and took a deep breath. It was the point when my lungs were full that I heard the ticking.

There is an explanation for the split second pause before I baled out. On every operation I carried a sharp pocket knife, which had a score of uses. On this occasion my hand flashed to my pocket. There were wires I might cut if I were quick enough. But ... my pocket knife was already performing one of those many functions for which it was famous – tied into the hauling cord to stop a knot slipping. There was only me against the bomb and *it* couldn't lose. And so I went, taking the most direct line to sanctuary, not paralysed but jet-propelled by fear.

The warrant officer's eyes met mine with an unspoken question as I broke the news about the bomb's astonishingly robust state of health.

'What do we do now?'

Here we were hiding behind cover with a device ticking merrily on top of the chimney. After more than two hours' work the damn thing still wasn't neutralised.

'Must've been the hard crack it got coming up against the ledge when it jammed in the chimney,' he said in his quiet Midland tones. 'That's what started it ticking, sir.'

We peered together at the mocking black pipe, with the fishing line and cord still attached to it, trailing across the tiles and falling limply to the ground and across the garden.

'I suppose I'd better go and give it another pull,' I said wearily.

'I suppose you'd better, sir.'

I was painfully conscious of the fact that if the fishing hooks were not secure in the string handle, or if the line broke again, the bomb could return whence it had come.

I broke cover, took up the end of the line and moved back to shelter.

'Well, here goes.'

The bomb responded to the tug immediately. Plunging from the chimney, it bounced crazily over the tiles, leapt the gutter and hit the ground. There was no bang. Cautious study of the scene showed the threatening clock had been thrown out of the top of the casing and lay white and gleaming some distance away. Whatever else made up the bomb was still inside the pipe.

So far so good, but neutralising bombs is only part of the ammunition technical officer's business. He also has to provide evidence for the police which, with luck, may put a terrorist behind bars for life. In the case of the chimney bomb, we needed the guts of the beast. So once again the hook and line set came into play, being attached firmly into the device. I ran the line through a pulley, which is part of the bomb team's kit, taking it behind a stench pipe running up the wall. From behind cover we pulled hard so the bomb crashed upside down against the side of the house shooting its contents onto the ground.

'Looks interesting,' said the warrant officer.

It was. I separated the contents of the device and called up the team. While they put the pieces into forensic bags to be dealt with later, I had them listed.

A final check showed that the plastic pipe had contained

15½lbs of CO-OP, a home-made mixture of weedkiller and nitro-benzene that smells strongly of almonds.

'If that'd gone off it would have blown the bungalow into the Irish Sea,' said one of the team thoughtfully.

With the explosive there was a 9-volt Exide battery, a short delay detonator and the alarm clock. It had been set to explode on the hour hand making contact with a wood screw and had two hours to go when we neutralisd it. When the remains of the bomb had been parcelled up I made a final search of the bungalow. Everything had been cleaned thoroughly so it was much easier to check that nothing else had been hidden inside.

What the devil was I doing, I asked myself as I poked here and there, looking for bombs in other people's houses? When I might be with my family in England or Germany with my only worry to make my bank account balance at the end of the month.

'Didn't you hurt yourself when you jumped off the roof, Colonel?' Coupe asked as we headed back to HQ through the dusk.

'Not a scratch.'

'Must have been the speed you were going. Didn't touch anything long enough to get a graze. Didn't know you could move so fast.'

'Didn't know myself, until today.'

'No, I don't suppose you did.'

We drove on in silence until the glare above the Maze, just like a fairground, told us we were nearing Lisburn.

At the vehicle check-point someone had just joined the Ramp Riders' Club. A sorry-looking Ford Escort was flopped across a raised barrier and the Chunkies and the driver were involved in the usual recriminations. As we edged past on the other lane one of the Pioneers, a coloured lad, beamed at us and winked before coming out with the current catch phrase:

'Good 'ere, init, mates?'

CHAPTER TWO

Hello to All That...

Bombs started to play an important part in my life when I was five. It was nearly a year after war broke out and we were in the middle of the Blitz. I say 'in the middle' deliberately because my family lived in London's dockland where my grandfather, Albert Patrick, was the landlord of the Rouel Tavern, Rouel Road, Bermondsey. My mother had looked after the pub since my father was called up and my twin brother and I got a very clear boy's-eye-view of what high explosive could do to bricks and mortar and all that went with it.

The first casualty that brought the gravity of events home to us was our dog. He was hit and killed by a piece of shrapnel. But worse followed soon afterwards. Because it backed onto the railway arches and promised greater security, we used to take shelter in another pub. the Lion in Enid Street, during raids. We emerged one morning and learned that a Nazi bomb had put paid to the Rouel Tavern and all our belongings.

The event didn't make a great impact on our way of life. My grandfather simply took another pub, The Rising Sun in Jamaica Road. This was a stout Victorian edifice which bore a charmed existence under the deluge of Hermans and other forms of hatred that showered around. Some of our customers were not so lucky as the pub. I have a vivid memory of three air raid wardens popping in for a quick drink one night in 1944. All were killed within five minutes of leaving the bar.

Equally vivid is an earlier memory of a blaze started when a stick of bombs fell 100 yards from The Lion and ripped open a gas main. I wouldn't see the like of those flames for 30 years.

Hitler's secret weapons also made their appearance while we were living at The Rising Sun. The first, the V1 Buzz Bomb or Doodlebug, was a pilotless aircraft filled with high explosive. Somehow, because you could see it, and because it was reckoned to be dangerous only when its motor cut out, it didn't worry us as much as it should have done. The V2, a straightforward rocket, was much more terrifying. It travelled at such a speed that it outstripped its own sound. Only after it had plunged to earth and exploded did one hear the dreadful roar of its passage through the air.

My brother and I viewed with awe the heap of rubble and plaster that had once been Dockhead Bend after it was hit by a V2. A convent school was among the many buildings which had vanished in the holocaust, and a weary rescue worker said in the pub that night, 'They've decided not to go on looking for bodies. They're in such small bits it's just a waste of time.' Hardened though we were, after this time, the remark struck a chill chord. But the luck of the Rising Sun held. When they sounded the last 'All Clear' the old pub was still there – 'an oasis in a sea of ruins,' my grandfather called it.

Those early impressions of the efficacy of high explosives never left me. Thirty years later when I was posted to Northern Ireland I reflected on the devastation I had seen in Bermondsey, with its shattered streets, flaring gas jets and obliterated buildings. Clearly Belfast's bombs would be neither as big nor as plentiful but, like Hitler's, they would destroy people and things with the same mindless savagery and merciless violence. My first impression of Belfast when I made a reconnaissance visit in January 1976 was that it looked like Bermondsey after the Blitz – run down, gaps in the skyline – but without the queues outside the shops. The queues were at the check-points to get into the city centre.

The Royal Army Ordnance Corps, in which I served after leaving Sandhurst, are the suppliers to the Army. If a unit needs rations or boots, or tanks or radio sets, it gets them from the RAOC. You can get a button – or a stand of Colours for an infantry battalion. In particular, you can get your ammunition from the RAOC.

The RAOC is full of specialists of one sort or another. It trains and employs the tradesmen who have to be conversant with the characteristics and intricacies of shells, bullets and similar goodies, so that they can be inspected, repaired and stored safely. In the case of non-commissioned officers and warrant officers they are known as ammunition technicians – ATs – and when they hold commissioned rank, as ammunition technical officers – ATOs (pronounced as in potatoes). It is from this reservoir of manpower that the Army draws people to deal with the terrorist bombs in Northern Ireland.

As usual where the Armed Services are concerned there are firm demarcation lines governing the handling of suspect explosive devices. The Royal Navy tackles objects below the high water line, as they have years of experience dealing with mines, naval shells, and other menaces which have been subjected to the action of salt water. On land, the Royal Engineers deal with Hitler's souvenirs, mainly because they have the required skills and can call on heavy equipment for digging and earthmoving. Anyone who watched the TV series 'Danger UXB' will have some idea of the depth at which aircraft bombs are found.

Members of my own Corps deal with our own ammunition, whether it is a Mills bomb which turns up on some old lady's mantelpiece where Dad left it on return from Flanders, or a modern 105 mm shell which, for some reason, fails to explode on the ranges. They also deal with terrorist-made 'explosive devices' or 'improvised explosive devices'.

Today all RAOC officers, NCOs and warrant officers in the ammunition 'trade' may expect to be called on to serve a period in Northern Ireland in an EOD unit. In my own case the call came after I had spent 23 years in the Service without filling a true combat role. It may be as well, therefore, to cast a questioning eye over my career.

I don't suppose I was the first boy from Bermondsey to enter Sandhurst direct, but I am pretty sure that if there were any others around in 1951 we could have held our get-togethers in a telephone kiosk with room for a guest speaker.

My father, who served in the rank of gunner during his war service in the Royal Artillery, tore up my papers when they first arrived but later gave his consent. The headmaster at St Olave's Grammar School, near Tower Bridge, told him he didn't think I was good for anything else.

And so off I went and discovered that perhaps my father had not been such a bad judge after all and that having been captain of boxing at St Olave's – it was essential to be useful with your fists when you had to walk home up Tooley Street wearing a straw boater – didn't really cut much ice among 850 cadets who came mainly from public schools.

If my true aim was to escape from my environment, then verily I had succeeded. The Army was trying to get back to peace-time soldiering and boys like me were not too common. I had to get used to being called a number of things, including 'the pub keeper's son' (my father was carrying on the family tradition) and the 'dirty East Ender', by some of my snootier contemporaries in Rhine Company, Victory College, Sandhurst.

The ordeal of coping with melon on my first dining-in night remains a ghastly memory to this day.

Fortunately my company commander, Major Bloomfield, of the Royal Engineers, told me regularly, 'I am commissioning you and not your father. Get on with it.' I passed out 21st in order of merit and carried off the Royal Army Ordnance Corps Sword and the Modern Subjects Prize. Perhaps we grammar school boys weren't too inferior after all. Ironically, after I was commissioned, the situation was often reversed. Some local people who had known me as a boy would cross the road to avoid speaking to me when I went home to the pub in uniform. On the other hand there were dockers who would never let me make the journey from Bermondsey to Liverpool Street in anything but one of their lorries when I went back off leave during my later service in Germany.

When I went home I kept out of the bar as much as possible. After all deductions had been made from my pay of 17s 6d a day (say 87 pence) I couldn't afford to buy a round of drinks. It helped to give me a reputation for being stuck up.

I don't deny that I felt it something of a let-down when I joined the Corps. Before I'd gone to Sandhurst I'd spent three months with the Queen's, where my maternal grandfather had been a sergeant in the First World War, and I'd enjoyed it. However, much as I may have wanted to serve in the Queen's Royal Regiment I doubt very much if I would have been acceptable in the Mess in those days. Instead, I decided to spend my time in a semi-technical Corps. When I joined the RAOC I consoled myself with the unlikely thought – for a man who had chosen soldiering as a profession – that at least it would give me an opportunity to study business management. Any opportunity for excitement or adventure I would regard as a bonus, never dreaming how long it would be before I collected it.

For 18 months after leaving Sandhurst I trained recruits in No 1 Regular Training Battalion at the ROAC Training Centre at Blackdown, Hants. This was followed by a three-year tour in Germany, half of which was spent as the mechanical transport officer of an ordnance field park, and half issuing vehicles and other spares to the 7th Armoured Division, then commanded by Major-General Shan Hackett.

I left the field park in 1958 to go on the 'long ammunition course' – six months at the Royal Military College of Science, Shrivenham, followed by six months at Bramley, near Basingstoke.

One thing which was not taught at Bramley was Explosive Ordnance Disposal. Although the Cyprus emergency was in full swing and some officers were dealing with devices, it was not regarded as an important line of business for a career soldier. It was looked on as a dubious function pushed in our direction by the Home Office, and was mentioned only in passing.

We were a motley crew on the ammunition course and the only officer who was sufficiently qualified academically (in my opinion) was removed for sheer idleness. He landed in gaol some years later for illicit arms dealing. The senior student was a pleasant enough officer nicknamed Creeping Jesus. We persuaded him to use irregular material when doing a practical experiment in applying nitrate to cotton, and regarded it as a great joke when he blew his apparatus to pieces.

The net result of the ammunition course was that I passed out with the august title of 'inspecting ordnance officer' – it was before the days of ATOs. The only other British lieutenant on the course, who qualified at the same time, preceded me in Northern Ireland as CATO.

For the next two and a half years I worked as a sort of sleuth for the ammunition inspectorate in Germany, carrying out investigations into defects and accidents. In addition, we removed and destroyed defective ammunition from units. I learned a lot about ammunition and even more about human nature. More than once I was called out because a live round had been inserted in a stock of practice ammunition issued for an exercise. Inevitably the result of this tampering would be the postponement of the exercise and some soldier, somewhere, would be able to keep a date with a *Fräulein* that was in jeopardy. Occasionally there were unpleasant accidents but the bodies, dead or otherwise, were always removed before we arrived. Some cases were more interesting than others. A Guardsman was suspected of attempted suicide by trying to blow himself to pieces with a 2 inch mortar bomb. I was able to show he had snagged the firing toggle of the weapon as he left his slit trench with it loaded – strictly against all rules. He was lucky to lose only part of his arm. In another case, an officer set up an unauthorised booby trap which went off in his face and cost him his sight.

Although I never came across one case of faulty ammunition causing an accident, the unpredictable nature of explosives in general was planted firmly in my mind.

Just at the point where I began to feel I was a real expert, the implacable system activated itself and I was posted to an appointment which had nothing to do with ammunition – a desk job involving the provision of spares for such things as earthmoving equipment, tanks and helicopters.

Eventually a two-year spell at the War Office was ordained where once again I commanded a desk dealing with RAOC policy matters – defence reviews, organisation, reorganisation and what at times seemed to be disorganisation. The tour did have its rewards. Apart from being back in my home town, I learned that I had been selected for a place at the Army Staff College. My qualifications in the ammunition business, however, required me to spend an initial 15 months at Shrivenham before going on to the year-long course at Camberley.

Having qualified at both places I spent two more years at a research and development establishment in Kent.

In due course the mandarins of the Military Secretary's department, who guide the stars of officers, sent me to command 44 Replenishment Park Company, part of 1 Combat Supplies Battalion in BAOR, where we handled everything from mines to guided missiles.

Finally, in 1974, I found myself a lieutenant-colonel in the holy of holies: the Directorate of Land Service Ammunition, at Didcot in Berkshire. Without having heard a shot fired in anger, I had become an ammunition egghead, dedicated to assessing the effectiveness of our ammunition stocks when used by troops armed with standard weapons and equipment.

Life at DLSA (pronounced Dilsa to the initiated) is rather

pleasant – it might approximate to that of a Varsity don in ammunition – with a comfortable mess, not over-populated, and a cuisine with an enviable reputation. Not for nothing are the Corps the suppliers to the Army . . . In the warmth of the lounge, after a good meal and a glass of wine, one could contemplate some of the nicer things in life. And, to be frank, service in Northern Ireland wasn't one of them. The plain fact of the matter is that by comparison with the major role of the Army as a vital part of the North Atlantic Treaty Organisation's forces in Europe, the Northern Ireland theatre is, in overall terms, very small beer indeed.

Senior officers see the commitment as an irritating nuisance, interfering with training in Germany, upsetting families in BAOR, and sometimes causing ill-feeling between those who have never served in the Province and those who have seen too much of it. Human nature being what it is, many of the people who have never done a tour over there do not hang on the words of those who have. It is only when individuals are due to make their first trip to Northern Ireland themselves that they discover, to their astonishment, that the stories from the Six Counties have an irresistible fascination after all.

So it was with me. Suddenly the corridors of DLSA lost some of their atmosphere of assured, cloistered calm. My interest in APDS* ammunition waned. A lifetime's accumulated knowledge of guns and the balancing of their tempered steel barrels, of the influence of sighting systems and the critical nature of charges, gradually assumed a relevance of an academic nature. True, it was more important in the long run to ensure that a shell from the 120mm gun of a 52-ton Chieftain tank would run true, penetrate and destroy a Russian T-54 or the T-72 (being talked of even then in respectful tones).

But it was the short run, and it was mundane things like the blast qualities of home-made explosive stuffed into a milk churn which increasingly began to absorb my attention once Brigadier Jimmy Lawrence-Archer, the Director of Land Services Ammunition, told me that in about a year's time I would be taking over as Chief Ammunition Technical Officer (CATO) in Northern Ireland. I was to replace my old

* APDS: 'armour piercing discarding sabot', applied to a type of shell for knocking out tanks.

companion of the 'long ammunition course' who had been at DLSA working on computer systems when I arrived there in 1974.

Belfast has a number of impressive thoroughfares, but the road from the ferry landing is not one of them. It passes through dirty brick Victorian streets not unlike London's dockland before the war, and to me, driving past on my way to Lisburn and Headquarters Northern Ireland, they looked sinister and mysterious.

It was a dull morning in June promising little. The overnight crossing had been uneventful and the security checks on arrival lacked portent. A number of passengers were coming back from holiday, I gathered, and I was struck by the warmth of feeling shown by individuals and groups waiting to welcome passengers beyond the security barriers. But these were personal and private gestures, and instinctively I knew there were things they knew and shared from which I was automatically excluded. Yet there I was, dedicated, pro-grammed, ordered, call it what you like, to do everything in my power to fight the evil that threatened these people.

The few folk about at that early hour looked alien, the faces of the stray pedestrians on the zebra crossings inscrutable. We passed a Land-Rover manned by unconcerned troops from a Gunner regiment and received a few bored glances.

Peter, the CATO I was relieving, had met me in a civilian vehicle and was in a hurry to get back to Lisburn.

'It's breakfast,' he explained as the car gathered speed. 'If we don't get there shortly they'll have scoffed the lot. By the time we've dumped your kit in your room and changed it'll all be gone, unless we put a spurt on.'

Occasionally we sped past stretches of cracked and blackened shop façades and boarded-up buildings. Peter drew my attention to walls with little white circles painted on them.

'They've been stuck up to indicate possible sniper positions,' he said. 'It pays to take a look at them. You never know ... you might be lucky enough to catch a glint of the barrel before you feel the pain.'

We turned past a pub with wire netting on the windows. A scorched decorative beer keg which had once held a clock hung drunkenly over the pavement. I was reminded of pictures of depressed areas in the North in the Thirties.

'I don't think we're going to starve after all,' said Peter as we negotiated a large roundabout overlooked by a tall grey building. As we flew up the motorway there were green fields on either side. Across them on the right a big cemetery and the roofs of a suburb were emerging from the light mist.

Peter answered my unspoken question. 'Andersonstown.'

In Lisburn, which lies just off the motorway, we ran into a slight traffic jam. Woolworth's had been bombed the previous month; the town centre was sealed off and had to be circumnavigated. The hold-up was only momentary, however, and we scuttled through the main check-point, down past the playing fields, in fine time.

After breakfast we paid a quick visit to the hut which housed the HQ of 321 EOD unit and received a rundown on the events of the night from the duty officer. From there we went to the Ops Room in the main headquarters, a modern red-brick building with an entrance hall like a cinema foyer, with blue-clad Ministry of Defence policemen checking passes instead of tickets.

Major-General David Young, the Commander Land Forces,* was particularly interested in Peter's report on the latest incident. A soldier in the Parachute Regiment had been killed at Crossmaglen by a bomb which had been hidden in the carrier on a parked bicycle. First indications were that it was a radio-controlled device. It came home to me quite starkly that from now on, as the troops say, things would be 'for real'.

Eighteen months had elapsed since I was first told that I was selected for the Northern Ireland job. It should have been a year, but there was some difficulty in finding a replacement for me at DLSA and so Peter had soldiered on till I could be released. During the interim period the thing which presented the greatest personal problem was when to tell my wife. In the end I put it off until there were only six months to go, and it became necessary to make certain obvious arrangements, like drawing up a will which, with optimism or through careless-ness, I had neglected to do up till then. With three growing children and a mortgage on a house in a pleasant old world village in Norfolk I didn't want to leave any untidy ends

* Later Lieut-General Sir David Young, GOC Scotland.

should the worst happen. I could just hear the old boys down at The Magpie saying in the event of my demise: ''Ear 'e left 'is affairs in a terrible mess, poor chap.' They get to know everything down at The Magpie in the end.

My wife, Pat, took the news in the way in which Army officers' wives are expected to take such news.

'I suppose it had to come some time,' she said. 'Just you be careful, that's all.'

I promised I would.

In a comparatively small force like the British Army, people in the same line of business get to know each other. So do their wives and families. And my wife had not forgotten about Bernard Calladene.

Bernard and I had known each other since 1954 when we went on the Ordnance Officers' Course, the statutory training given to commissioned members of the RAOC after they leave Sandhurst. Our paths had coincided or crossed regularly throughout our careers. We had served together during the two years I spent at the Ministry of Defence and he had commanded 44 Replenishment Park Company, RAOC, in Germany, which I took over at a later stage.

A dark-haired man, well over six feet tall and broad with it, he had a generous nature to go with his physique. He was killed one rainy night in March 1972 while approaching a suspect car in Wellington Street, Belfast, not far from the Europa Hotel. It blew up as, torch in hand, he walked towards it. The OC of 321 EOD at the time, Bernard's death was a great loss. He left a widow and young children.

Captain Barry Gritten was someone my wife and I had known when he was Adjutant of 1 Combat Supplies Battalion. He too was married with a family when he was posted to Northern Ireland. One day down country he went into a store which contained a pile of suspect fertiliser bags. He did not survive the explosion. His wife was expecting another child at the time.

Gus Garside, a stout-hearted, cheerful warrant officer who served under me at DLSA for 18 months, was another victim of the bombers, killed going through a hedge some distance from where a suspect device had been reported.

Like most soldiers bound for a tour in Northern Ireland, I imagine, I expressed the view outwardly that the odds were against anything happening to me – as indeed they were.

Inwardly, however, I acknowledged that if big, calm, sensible Bernard could fall a victim to the bombers, I could count on no special act of Providence to protect me or the men under my command. The death rate among operators at the time I was due to go to Northern Ireland was five per cent. Roughly translated this meant that I could expect one operator at least to be killed during my tour. Which one?

Survival depends on a number of things – the intensity of violence in the Province during a tour, the extent to which training has been absorbed, experience, judgment and temperament. And it is a common misconception that all the men who deal with suspect terrorist devices are volunteers. This has not been the case for some years, whatever may have happened in the dim and distant past. As I said earlier, any qualified (ie rated Class I) senior rank or warrant officer or officer in the ammunition trade can be called on. The way to handle explosive ordnance devices is studied during training. It is supposed to be just a normal part of their job. But what is normal about it? I do not belong to the school whose motto is, 'If you can't take a joke you shouldn't join,' and the Army also recognises that temperament plays an important part of the makeup of an EOD operator. Personnel selected for posting to Northern Ireland have to undergo two processes.

The first is carried out by a psychiatrist, and is called 'psychometric testing'. The object of this is to find out whether you are normal and whether you can be relied on to react normally under stress. In my case I was asked such things as, 'What are your favourite colours? Do you like looking at railway lines disappearing into the distance? Do you enjoy sexual dreams?' Following the session with the psychiatrist people are graded, although they are never told the level they achieve. As far as I am aware, the only positive thing psychometric testing does is to identify the clinically insane.

It struck me at the time that it was strange to be tested after I had been posted as any lieutenant-colonel who fails the test ought to be sacked immediately, and certainly removed from command. I gather the reason was that I was so long in the tooth, psychometric testing had not been part of the system when I was learning the business. Young officers nowadays, I'm pleased to say, are checked out before the taxpayers' money is spent on putting them through the ATO course.

. For the other ranks it is a slightly different matter. They start to train at eighteen and have several years to mature before becoming sergeants and warrant officers in the ammunition trade. Any signs of unsuitability should have emerged before they reach the stage of selection for Northern Ireland duty, but in any case they go through the psychometric testing procedure before they begin their tour.

Having cleared that hurdle the operators then undergo a three-week course designed to assess their mechanical skill and their ability to think through and work through an EOD situation in a safe and reliable manner. After that it is up to the brigadier at Didcot, the Director Land Service Ammunition. He alone has the power to accept or reject the results of the psychometric tests or, if the occasion arises, the pre-ops course.

Being the arbiter in these cases subjected the brigadier to considerable pressure, especially where senior ranks and warrant officers were concerned. To decide that they were unsuitable, for whatever reason, almost certainly meant the end of their career in the ammuniton trade. They could be required to remuster as stores NCOs or in a clerical grade, at a much lower rate of pay ... not to mention the damage to their self-esteem and their loss of prestige among their fellows. Borderline cases were, therefore, a real problem.

The one person who was not consulted about doubtfuls, at least in my day, was CATO Northern Ireland. Fortunately for me, while I was in command there, I could rely on a very experienced officer to look after my interests, Lieutenant-Colonel Mike Newcombe, then serving at Didcot on the Staff of DLSA. Mike had been awarded the George Medal for bravery while major in command of 321 EOD unit in Northern Ireland and knew the score. He and the first-class team of officers working with him would always do their best to let me know if they thought I was about to be presented with a problem. They fixed things so that the balance between the good and the not-so-good in any one section was always about right, risking the wrath of their seniors in some cases in order to achieve their ends.

Despite the precautions, however, men did slip through the net. Psychometric testing may have established that we were not mad but it did not reveal everything. The danger of employing an individual who is not mentally attuned to the task was brought home forcibly one day when I was told by the ops staff at 39 Brigade that one of my operators had been injured in an incident in Queen Victoria Street, Belfast. Immediately I sent for Corporal Coupe and we drove smartly down the motorway from Lisburn, picking up the news by radio en route that the Incident Control Post (ICP) was on a corner near the BBC's offices.

The bomb team's vehicles were still across the road when we got there and the other end of the street was also cordoned off. The ATO, a sergeant, had been taken to hospital, along with the soldier who had been acting as his escort, but the team's No 2, a good, reliable ROAC corporal driver, told me what had happened.

Armed men had forced their way into a shop in Queen Victoria Street and placed an object in the drawer of a piece of furniture. After the team arrived the sergeant in charge had moved up the side of the street opposite the shop and placed his escort in a doorway so he could see along the road and cover him. He stepped off the pavement to cross the street directly opposite the shop when the bomb detonated. From twenty yards he was caught by the blast, the remains of the plate glass window, and a lot of debris. His escort had also been hurt, but by some lucky chance, not seriously.

Those were the cold facts and, on the face of it, they seemed to suggest that everything was in order. I went up to the shop with the No 2 and saw for myself. I didn't bother to ask any more questions and, as my first concern was to arrange for the area to be cleared – ie, pronounced free from danger, I called up another team from the City Centre to relieve the reduced section still in Queen Victoria Street.

After that, my main consideration was for my ATO. How badly was he injured? How long would he be in hospital? Was a replacement required?

'I'll have to chat up the sisters,' I remember telling Coupe as we drove to hospital. 'We can't afford to be without an ATO for any length of time.'

It turned out that there was not much point in 'chatting up' the Queen Alexandra's Royal Army Nursing Corps. The sergeant had been badly cut on the wrist and had various other lacerations on both arms I was pleased to see that he had no damage to his face or the rest of his body. The staff at the Musgrave Park Hospital said he would probably be released for light duties in two or three days and, on my return to headquarters, I arranged for a sergeant clerk, who was also an ammunition technician, to leave his desk at Lisburn and take over the team in Belfast for a spell.

It was through him that I learned the full story. As the injured sergeant's condition improved and he was almost fit to take over the team again, the No 2, the reliable corporal I mentioned earlier, told the replacement sergeant that he didn't relish the thought.

'I don't want to work with him again,' he said. 'He's too dangerous. He's going to get himself killed and probably someone with him.'

On hearing this I asked the OC of 321 EOD Unit to interview the No 2. He extracted the information that the sergeant was in the habit of making a manual approach to each and every device he was called upon to deal with. Furthermore he did not allow a 'soak' period – when the suspect object or the target is kept under observation but no one goes near it – and he did not wear a protective bomb suit. On the morning he was injured he had arrived at the ICP, set up his vehicles in a protective screen and run out Wheelbarrow – the excellent little robot machine which, by remote control, can do a variety of jobs without risking an operator's life. On this occasion, however, and apparently in accordance with his usual practice, Wheelbarrow was simply left by the side of the vehicle which had carried it. The sergeant just walked up to the bomb. He did not wear his bomb suit – though he did put on the helmet and pull down the visor – and, because the weather was warm, he even had his sleeves rolled up.

Apart from breaking all the rules, the sergeant had taken risks which were worse than stupid and reckless. In leading his escort into the danger zone so soon after the device had been reported and before the required drills had been carried out, he had been guilty of an act which can be regarded only as criminal.

While I was studying the report on the case, I noticed the sergeant's parent unit in Great Britain and something clicked. Some time previously, a captain serving as an ATO at Lurgan had told me he had doubts about an NCO from his previous unit who was due for an emergency tour of Northern Ireland. The captain, who was about to return home, had said bluntly, 'In my opinion, Colonel, I don't think he is suitable for this job.'

'Why?' I asked.

'He's young, his marriage has broken up, and I don't believe he is capable of reacting sensibly under strain. The break-up has been too much for him.'

'Did you mention this to anyone before you came out here?'

'Yes, I told my OC and I mentioned it when I was at Didcot.'

As far as I was concerned, at the time, that was that. The powers that be had been told – the people responsible for the selection process. I could not see any way in which they would allow an unstable NCO to serve an emergency tour in Northern Ireland.

Further investigation, however, revealed that this was the very man I had been warned about by the captain from Lurgan. I interviewed him myself and he admitted that since his marital troubles had begun he didn't really care what happened to him. Further psychometric tests showed that he had what amounted to a death wish. His personal problems had assumed greater proportions than those presented by the bombs he faced. Clearly the man was a menace while he was in this state of mind and he was sent back to his parent unit in Great Britain the next day – with my recommendation that he should return to complete an operational tour once he had sorted himself out. This, I understand, he eventually did.

What remained disturbing was the fact that somehow Didcot, despite warnings and despite the testing, did not seem to be aware that they had sent out an NCO whose mental state would put him and his team at risk. If later I began to have a bee in my bonnet about the selection of operators, The Case of the Sergeant Who Slipped Through the Net was one of the reasons for it.

CHAPTER THREE

Murder by Clothes Peg

One of the most inhibiting factors about the military campaign in Northern Ireland is its artificial nature. There are too many restrictions on the Security Forces which bear no relationship to the realities of combatting guerilla warfare. The denial of the right to pursue raiders across the border, even though they may have committed murder, is only one of them. As far as the Army is concerned, no soldier is allowed to leave the board on which this dirty game is being played. Their opponents are able not only to leave the board but crawl under the table, switch out the light, and leave the room if they feel like it. Although in some ways, the confined area of operations has exerted an atrophying influence on the thinking of the urban terrorist.

In geographical terms, Northern Ireland is small. Its 5,000 square miles equal the size of the old county of Yorkshire. The population is 1,500,000, around twice that of Manchester or about a fifth of London's.

There have been occasional sensational attacks on mainland Britain, and prominent individuals have been assassinated – the British ambassador in Dublin, Mr Ewart Biggs; Mr Airey Neave, at the time Tory spokesman on Northern Ireland; Earl Mountbatten of Burma while on holiday in Eire – but since the current troubles began in 1969, Republican terrorists have mainly concentrated their efforts within the Six Counties.

Yet, after all their exertions, they cannot claim any territorial gains – the proof of success exhibited by insurgents in classic 'wars of liberation'. Nor can they point to any social or political advantages resulting from their applied violence.

In most democratic countries, elections are held to decide who is to govern. In Northern Ireland, elections have been held to resolve the question of how the country is to be governed at all. Since March 1973, the people of the Six Counties have gone to the polls a number of times. They have been asked to vote on everything from local government to the territorial sovereignty of the British Isles.

The political situation remains a stalemate. The British government has been unable to overcome the basic fact that the Loyalist population has an inbuilt electoral majority and are not prepared to concede to the mainly Catholic Social Democratic Labour Party a place in the government as a right. No amount of argument by Republican politicians on both sides of the border can obscure the fact that though the minority in Northern Ireland is Catholic the minority in the geographical entity of Ireland is Protestant.

The terrorists' policy of driving the Army out and coercing the people into submission has proved to be sterile. Their obduracy and fanaticism is mirrored by extremists of a different persuasion.

Incapable of finding or devising political stratagems or a peace strategy, unwilling to seek means of compromise, terrorist leaders inevitably return in their frustration to the line which requires the least thought – that of intimidation. To disguise the futility of their arguments, to bind wavering followers to them, to impress potential followers and rally sympathisers, they have to make it appear that they have something new in their armoury – just as Hitler constantly fell back on promises and displays of secret weapons to avoid defeat. As the terrorists have nothing new, they simply vary established techniques.

At moments when, as *The Irish Times* once said, 'the air is thick with impossibilities,' the noise of a terrorist bomb will be heard sympathetically at least by the people who planted it.

To get the work of military EOD squads into perspective, it is necessary to look at the evolution of the Army's role in Northern Ireland over the years. It commenced on Thursday 14 August, 1969. After several days rioting in Londonderry it became clear that the Royal Ulster Constabulary and the B-Specials were no longer able to cope with the situation. On that date men of the 1st Prince of Wales's Own Regiment of Yorkshire (an amalgamation of what used to be the West Yorkshire Regiment and the East Yorkshire Regiment) moved in as a peacekeeping force to back up the police.

Other troops moved into Belfast, separating the opposing mobs. To begin with the soldiers were welcomed by both sides, the Catholics seeing them as a force able to shield them from Protestant thugs and prepared to deal with them fairly in place of the RUC and the B-Specials who were regarded with suspicion and distrust in many areas. (The Hunt Report recommended the disbandment of the Specials in October 1969, a proposal which was implemented the following spring.) For the Protestant and the uncommitted members of the public, the troops were representatives of the Crown employed on behalf of all to subdue the turbulence which threatened to get out of control. Strange to recall, housewives in Republican areas, such as the Falls Road and the Ballymurphy Estate, offered cups of tea and pieces of cake in those days. Teenage Catholic girls packed the discos run by troops in Belfast.

Within a matter of months the same wives were banging dustbin lids and screaming insults at Army patrols moving through the estates and the girls were carrying ammunition for gunmen prepared to assassinate their erstwhile dancing partners.

Sections of the Loyalist community were equally resentful of the Army's presence, particularly when troops were used to enforce a ban on marches. A massive riot in the Protestant Shankill Road area in Ocotober 1969 culminated in the shooting of a policeman. The troops were ordered to open fire and two civilians died.

As disorder increased more men arrived and a naval vessel, HMS Maidstone, was anchored in Belfast harbour to provide extra accommodation for the growing garrison. CS gas was used to disperse mobs and rioters retaliated by throwing petrol bombs.

The sorry situation was ideally suited to the purposes of Republican terrorist and the Provisional Wing of the Irish Republican Army in particular, and they did not hesitate to exploit it to the full. To them England's danger was, as ever, Ireland's opportunity.

A search of houses in the New Lodge and Falls Road areas of Belfast in July 1970 produced 95 illegal guns, most of them pistols and rifles but including eight automatic weapons. As well as a large amount of ammunition for the small arms, 100 bombs of various types and 250 lbs of high explosive were found.

The use of explosive devices developed slowly at first. Petrol bombs, sometimes referred to as Molotov cocktails, were common currency during riots in 1970, to the extent that the then General Officer Commanding Northern Ireland, Lieutenant-General Sir Ian Freeland, warned that youths carrying them would be shot. This followed a violent demonstration in the Ballymurphy and Springfield Road areas in which many soldiers were hurt.

Only nine explosions were recorded in 1969, but by 15 September the following year, the 100th blast had been registered, and by the end of 1970 the total stood at 153 for the twelve months.

This figure soared the following year as the violence worsened. In February 1971 the first soldier was shot and killed – Gunner Robert Curtis, the victim of a burst of automatic fire on the New Lodge Road. Five BBC technicians were blown to pieces by a landmine while on their way to service the TV mast at Brougha Mountain, County Tyrone. Three young soldiers of the Royal Highland Fusiliers were lured from a public house and murdered in a country lane near Ligoniel, Co Antrim, not far from Aldergrove airport, in March. On 9 August the government introduced internment without trial for suspected terrorists, and a fresh wave of bombings and shootings occurred. Arson became commonplace in parts of Belfast.

The bombers began to be more selective. The Europa Hotel in Victoria Street, unofficial headquarters of the Press corps covering the troubles, was among the targets. A police inspector was killed when Chichester Park RUC station was blown up for the fifth time. Fifteen people died when a terrorist bomb blew up in McGurk's Bar in North Queen St, Belfast. Most of the victims were Catholics and it was generally believed the explosion was caused by an IRA bomb going off prematurely while en route to a target.

Visitors to wounded soldiers sometimes found themselves placing their bunches of grapes or flowers next to a jar of surgical spirit on the bedside table – a jar which contained the fragments extracted from injuries. Home-made bombs stuffed with nails had become a familiar hazard for patrolling troops. Lengths of metal piping filled with explosives were another crude weapon in general use.

Far from crude were the sniper weapons emerging ...
American .233 calibre Armalite rifles and Eastern European
7.62mm Kalashnikov assault rifles among them. The IRA
even had a song about 'My Little Armalite'. For many service-
men, it is the high velocity bullet of such weapons that holds
the greatest fear in Northern Ireland. Unlike the GS (gun
shot) wound of the old .303 Lee-Enfield and similar weapons,
it travels at such a speed that it destroys the body tissue as it
passes through a victim. The shock effect is enormous and has
given a new phrase to military medical jargon – bullet trauma.

Disorder reached a peak in 1972 which has not been seen
since. On Sunday 30 January, men of the 1st Battalion, The
Parachute Regiment, were fired on when crowds gathered for
an illegal march in Londonderry. They fired back and killed
13 men, seven of them teenagers. An official inquiry was held
and the Lord Chief Justice, Lord Widgery, declared that there
was no proof that any of the dead men had been shot while
handling a gun or a bomb.

At the end of March the Northern Ireland Parliament at
Stormont was dissolved and direct rule from Westminster was
instituted. This made little difference to the terrorists. The Provisional IRA called for a ceasefire in June but carried on with
its policy of intimidation. Nine civilians died in a concerted
bomb attack on the bus depot in Oxford Street, Belfast, on
Friday 21 July, and the British public were shaken by the
ghastly sights shown on BBC TV that night. The incident recalled the lines from Edmund Blunden's poem *Escape* describing a scene on the Somme in 1916:

> Now God befriend me
> The next word not send me
> To view those ravished trunks
> And hips and blackened hunks.

The same month saw the 100th soldier killed in Northern
Ireland – Private James Joseph Jones of the King's Regiment.

A week after his death the Army launched Operation
Motorman, and swept away the so-called no-go areas, such as
the Bogside in Londonderry and Andersonstown in Belfast.
In Derry alone 25,000 lbs of explosives were unearthed.

At the end of a grim year the total of deaths due to terrorist action and its consequences reached 482. Of these 336 were civilians, the rest being members of the Security Forces. Nearly 4,000 members of the public were injured during this period, a large number by bombs (which caused 1,382 explosions).

The following year saw a falling-off in the level of violence – but only by comparison with the appalling toll of the 12 months which had gone before. In March three soldiers were lured to a flat in the Antrim Road, Belfast, and murdered. Gun battles involving Protestant terrorists broke out in the Shankill Road area in June and the 200th soldier died – Guardsman David Roberts, killed at Crossmaglen, in County Armagh, on the border with the Republic. Explosions were not far off the 1,000 mark.

A power-sharing excutive was created in 1974, with representatives from the opposing factions sitting together, but it did not last long. The Loyalist-controlled Ulster Workers Council brought Northern Ireland to a halt with a strike and forced the resignation of Brian Faulkner and other members. Attacks on the Army fell during this period and a total of 28 soldiers were killed compared with 58 the previous year. Bombings were down too, 685 recorded.

In 1975 the general level of terrorism and violence had diminished still further with the total number of deaths of members of the Army, Ulster Defence Regiment and Royal Ulster Constabulary totalling 31, the lowest figure since the arrival of the Army in Northern Ireland. Not since 1970 had so few explosions taken place – 399.

The number of sectarian killings soared alarmingly, however, and there were more than 140 known victims. The pattern of strife had changed significantly over the years, with riots now the exception where they had been the rule, particularly in Belfast and the city of Londonderry. Soldiers who had not served in the theatre since 1972 or '73 were struck by the different atmosphere. Criminal violence had attained a level of sophistication, but was sadly none the less hideous in reality. The terrorists, mainly the Provisional IRA, had switched to specific rather than general targets.

The change in emphasis was noticed dramatically in the border area of South Armagh. There County Monaghan in the

Irish Republic pushes into the north towards Lough Neagh creating a pronounced salient. From relatively safe bases in this region, raiders, either singly or in groups, have been able to cross the twisting and turning border with little difficulty. One result has been the high toll of victims in what the Press have christened Murder Triangle at the base of which lies Crossmaglen. Of that unquiet little market town, more later.

In the winter of 1975–76, activity on the border became intense, culminating in the Kings Mills Massacre on Friday 6 January, when ten Protestants were shot down and killed after being forced off a bus.

In February, Frank Stagg, a convicted IRA man, died as a result of a hunger strike and inspired three days of rioting in which seven people died and twenty-seven were injured.

About the same time, Mr Merlyn Rees, Secretary of State for Northern Ireland, confirmed his security policy to be reliance on the law and stressed that the police would play the leading role in enforcing it. He also announced the ending of Special Category status which meant that men convicted on terrorist offences after 1 March would be treated as common criminals and have no special privileges. This provoked demonstrations from the Loyalist groups in February, in opposition to the introduction of a conditional release scheme in connection with the ending of the special category status. The Republicans indulged in an outburst of violence in April objecting to the loss of what they claimed to be 'prisoner-of-war' status.

For a period there was mounting concern and men of the Special Air Service Regiment were sent to South Armagh at the beginning of the year, to be followed by the deployment of extra infantry units on the border. Although the spring unrest saw many bomb attacks and hoaxes and the hi-jacking of more than 100 vehicles, the Provisional IRA suffered serious losses and were forced to think again. Their active service units (ASUs) went to ground or began to move north to Tyrone and County Londonderry. Policemen, prison officers, soldiers and men of the Ulster Defence Regiment became prime targets. In retaliation, Loyalist extremists in the Ulster Volunteer Force, attacked selected Catholics in mid-Ulster and Belfast, killing six and injuring forty-seven.

All in all, new shades of ugliness were developing in the

summer of 1976 as the terrorists responded to successes by the Security Forces. The era of the booby trap bomb was dawning.

A variety of bombs were being used by terrorists at the time, but generally speaking the fillings were in three main categories – Commercial, CO–OP and ANFO.

Commercial, as its name implies, was standard blasting explosive of the type that is used in quarries. It blows up when activated directly by a detonator. Irish Industrial Explosives Ltd, in the Republic, were licensed to produce it and at the beginning of the campaign, plants making Commercial were an important source of supply for terrorists. Later the efforts of the Gardai in the Republic had a marked effect on this traffic.

CO-OP was essentially a home-made explosive, in which weedkiller (sodium chlorate) was mixed with nitro-benzene. Not as powerful as Commercial it is still highly effective. Smugglers have been known to bring it across the border in plastic 'sausages' weighing around 5 lbs. As with Commercial it responds directly to a detonator. Another home-made substance, ANFO, consisted of a mix of fertiliser (ammonium nitrate) and fuel oil. Because of its reduced power, by comparison with other explosives, it was used in large quantities. As much as 200 lbs of it has been crammed into cars while more than 1,000 lbs has been hidden in culverts which could then be blown up as a Security Forces vehicle passed over or alongside. An intermediary is needed to make ANFO explode, and CO-OP and Commerical are often used for this purpose.

Terrorists had no difficulty in getting hold of detonators at the start of the campaign, and obtained them by theft or through illicit deals and purchases abroad. Supplies of electric detonators were procured by one means or another from Irish Industrial Explosives Ltd or, in the case of Dupont detonators, from the United States.

Although the use by terrorists of manufactured detonators has been widespread for many years, it was only towards the end of my tour that an effective system of tracing them was introduced. The country of origin could then be identified. Commercial explosives from the Republic also had markings on them by that time. Some bombers made their own detonators, by the way, using mercury fulminate or zinc fulminate.

The early timing devices produced by the terrorists were often primitive. Some resembled contraptions associated with

35

the anarchists of the last century – alarm clocks with pieces of wire fixed to their winders so they caused an electrical contact at a fixed time.Later the IRA went into mass production of timing and power units (TPUs), churning out wooden boxes with five-ply sides, a three-ply base and a hardwood top. Inside it was normal to find a 60 or 120-minute pocket reminder-timer, two 1.5v dry batteries, and a clothes peg switch which had a brass strip or a drawing pin on each of the jaws, which were held apart by a dowel pin. Once the pin is withdrawn the jaws snap, the metal makes contact and the bomb is armed.

(A more unpredictable technique sometimes used was to hold the jaws of the peg open with a rubber band or a piece of soldering wire which was bound to give way at some time, preferably, from the bomber's point of view, when he had got rid of his device.)

The ready availability of these TPUs made it much simpler for the DIY bomber. All he had to do was to connect the detonator to the timer, insert the detonator in the explosive, set the timer and remove the dowel pin so that the bomb was armed. In the case of some boobytrap bombs it was left to the victim to perform some action that resulted in the dowel being pulled from the jaws of the clothes peg, the aim being instantaneous detonation. Just to help the bomber to practice, the IRA thoughtfully wired a test bulb into the circuit.

Cordtex, another commercial product which as its name suggests is an explosive in the shape of a piece of cord, was also circulating in some quantity and made a useful intermediary for the home-made bombs containing, for example, fertiliser. It is powerful enough to fell a tree encircled with it.

The other incendiary device in common use was concealed in a cassette which contained a mixture of sugar and some other volatile product. A wrist watch was used to delay initiation and the bomb burst into flame when an electric current passing through a wire had made it hot enough to get a reaction. Occasionally Loyalist terrorists used incendiaries in which the sugar and chemical mix was initiated by acid but these were not common.

One characteristic of fire-bomb raids was the tendency for the terrorists to use the incendiaries en masse, as they did in Ballymena on a Saturday in October 1976.

The home of the Rev Mr Ian Paisley, Ballymena is a town with atmosphere – vaguely threatening and a Loyalist stronghold. A neutral visitor would not be mistaken necessarily if he felt he was being watched. As an unfanatical Englishman once put it in the mess at Lisburn – 'I always feel like a Catholic in Ballymena – and a Protestant in Dungiven' – Dungiven being passionately Republican.

That day it had been my intention to go not to Ballymena but to Londonderry. However, when I reported to the Ops room after finishing my work, the GSO 2, Major Pat Moore of the Royal Scots, suggested it might be a good idea if I dropped in on the Security Forces there. Something seemed to be happening, though what exactly was not clear. Rather casually I said I would stop off on the way and, with Coupe driving, set off wearing a sports jacket and flannels in anticipation of a fairly relaxed weekend.

The RUC station staff at Ballymena were very friendly after I identified myself but were not demonstrably aware of what was happening.

At lunchtime a bomb had gone off in a Toyota car in a car park at Fairhill, Ballymena and four men in it had been seriously hurt. Three of the injured came from Bellaghy, Co Londonderry, from time to time the scene of some ugly incidents, and the fourth was from another troubled area. It was a reasonable assumption that they had been on their way to plant a device when it had exploded prematurely.

Not long afterwards, a report came in from the Security Forces base at Magherafelt, another hot-spot, saying they had received a warning that 30 bombs had been planted in Ballymena and were due to go off 20 minutes after the delivery of the warning. The police said that the ATO from Magherafelt, Sergeant B—, had arrived but they didn't know where he was exactly.

Coupe and I set off in search of this forlorn hope and his team, beginning at the car park where the trouble had started. The Toyota was there, under guard, but there was no sign of the sergeant. Finally we tracked him down to a hardware shop where he was just starting on a device. Apparently he had been clearing the Toyota when he heard of the other threats. In fact, just after one o'clock a bomb had exploded in a shop in Bridge Street, killing a young woman and causing a severe fire.

The only way for us to deal with an attack on this scale was for us to go our own way – but before we split up I helped B—, who was fairly new to the Province, to deal with the bomb in the hardware shop. It was a blast incendiary and we disrupted it with a charge which neutralised the explosive element of the bomb but blew creosote all over the place, a tin of the latter having being attached to add to the blaze.

Later it became clear that the premature explosion in the car had interrupted a serious fire raid by the Provisionals. As other devices were found and plotted on a map we saw what had happened. The bombers, who must have been assisted by young girls at some point, had gone on a round tour of the main shopping centre, carrying duffel bags or cassettes which they managed to hide in shops. Boutiques and fashion shops figured prominently as targets but a number of devices were also reported in Centuripe Avenue, leading to St Patrick's Barracks, depot of the Royal Irish Rangers.

With only my penknife as a tool I was not exactly equipped to deal with anything too sophisticated and when I got to a boutique called Top Gear I was quite pleased to discover that some rash person had removed a cassette which had gone off in the street. An ill-advised act to say the least – cassettes can inflict very bad burns – but all that was left for me to do was to collect the evidence for forensic purposes.

The bombs on the road leading to the barracks were also cassettes, and might well have been placed there to draw attention from the main attack on the town. One of them had gone off but four more had been placed in hedgerows by the married quarters. I managed to deal with them, mainly through judicious use of my penknife.

Meanwhile, Sergeant B— had been sorting out other devices in shops. I met up with him at the car park about teatime and decided to clear the car in which the four men had been hurt, this time borrowing his bomb suit to do the job. Three more blast incendiaries came to light, all of them in duffel bags and all destined originally for other targets. Basically they each held a pound of commercial explosive, a TPU, an electric detonator and a gallon of inflammable liquid.

A total of eighteen devices were planted in Ballymena that day (and there were others at Moneymore not far away). Between us we had been able to neutralise most of them and it was very bad luck that the woman had been killed before the particular device – a blast incendiary – was found.

Coupe and I carried on to Londonderry after completing the last task and Sergeant B— continued assessing the damage that had been done by our disruptive action, an official requirement that kept him busy until well into Sunday.

The major commanding 321 EOD unit who had to sign B—'s incident report put his own dry comment at the bottom of the sergeant's lengthy account of his operations: 'Quite a day!!' The sergeant would no doubt have agreed.

CHAPTER FOUR

Growing Pains

Like Topsy, the RAOC's organisation to combat the bombers in Northern Ireland just grew. In the begining there was only one EOD unit in the place, headed by a captain who had a warrant officer as his assistant. In the autumn of 1970, as the campaign gained momentum, it was recognised that reinforcements were needed to enable the Corps to carry out its role—it was, of course, still responsible for all the supply arrangements for the growing garrison and for housing the troops as well as dealing with the bomb threat.

Under the reorganisation, Major George Styles, who had been posted to Northern Ireland in 1969 as Deputy Assistant Director of Ordnance Services (DADOS), became Senior Ammunition Technical Officer (SATO). As a captain George had served on attachment from the RAOC in Malaya during the emergency. He had also been Senior Ammunition Technical Officer in Eastern Command, which brought him into contact with numerous problems relating to wartime bombs and shells. His knowledge of terrorists and explosives was invaluable. Twice in a few days in October 1971 George neutralised dangerous new-style bombs in the Europa Hotel, Belfast without any of the sophisticated equipment or knowledge that we had, and was decorated with the George Cross for gallantry.

George, who left the Province early in 1972, and retired as a lieutenant-colonel in 1974, was one of the first in this particularly challenging field. He and the men who worked alongside him had to learn the hard facts on the streets and bleak hillsides, building up their organisation as the campaign developed. Equipment, methods, people, attitudes, all change with the years. So do the numbers employed. My appointment as CATO required me to fill a number of roles. In a strictly military sense the responsibilities were straightforward. My primary function was to command all bomb clearing operations in Northern Ireland, through the agency of 321 Explosive Ordnance Disposal Unit which provided the operators and was commanded by a major. On my arrival the OC was a burly, cheerful Scot who was dedicated to pipe music, indeed reputed to play the pipes. He gave a good demonstration three months later on a set that he borrowed from the UDR.

Next I was adviser to the Commander Land Forces and, either through him or directly, to the General Officer Commanding and the Secretary of State, on all devices which appeared in the Province, their make-up and anything significant about the way in which they were being used.

On behalf of the GOC Northern Ireland, I was also the ultimate authority on whether devices should be neutralised or destroyed *in situ*.

In addition I had to provide intelligence information (which in some cases might be circulated to our allies), help conduct trials on behalf of Ministry of Defence research establishments and, in addition, feed back information to the training organisations in England.

It goes without saying that the day-to-day conduct of operations brought me into frequent contact with the Royal Ulster Constabulary at senior officer level and I had a close and essential liaison with the Department of Industrial and Forensic Science—known professionally as DIFS. Indeed, I had two warrant officers located with DIFS permanently, one of whom was kept busy dealing with all the detonators recovered (or their remains) and their indentification. This service was provided in the interest of the Home Office.

Another of my responsibilities was the Northern Ireland Ammunition Inspectorate. Just as in Germany or in Britain, we checked the scales of ammunition held by the Regular formations (which included the Ulster Defence Regiment), the Territorial Army and Cadets. And we checked storage and serviceability. As a number of units were always moving in and out of the Province in the four-month-tour cycle this task absorbed a very small staff of experts permanently. They had to undertake EOD clearance operations as well. Finally I was available to neutralise bombs myself.

The main strength of the personnel involved in bomb work

was located at Belfast, Londonderry and Lurgan at that time. At each place, one team was always on immediate readiness, fully dressed with their boots on day and night. The moment they were called out the next team moved up to the immediate state. The teams themselves, organised into sections, were allocated to brigades—the 39th Infantry Brigade covering Belfast, the 3rd covering the southern border and the 8th responsible for Londonderry and parts of Antrim and Tyrone. An EOD team consisted of an operator—the ATO—who was a senior rank, his No 2, generally a corporal, a driver and an armed escort. There was an officer who was also a qualified ATO commanding each section in each brigade. He had his own team as well, and was the adviser to the Brigade Commander on EOD support within the brigade area.

The three sections were organised differently due to the differing roles of the brigades. 39 Brigade had all its EOD section locations in the Belfast city centre. To cover the brigade rural areas I provided a fifth team, code-named Echo, from the staff at HQNI, HQ 321 and the Ammunition Inspectorate.

The Belfast section of four teams—call signs Alpha, Bravo, Charlie and Delta—gave adequate cover against most eventualities and enabled the troops to rest and maintain their equipment, do their laundry and write letters.

In 3 Brigade, two teams were located in the knicker factory at Lurgan and three more with the units stationed at Armagh, Omagh and Bessbrook, where they were on duty 24 hours a day.

The section in Londonderry was concentrated in a similar fashion to that in Belfast, but because the Foyle flows through the city 8 Brigade kept one team at immediate readiness in the city centre and, depending on the amount of terrorist activity in rural areas, a team was located at Magherafelt.

Unless specially authorised, teams operated in their own brigade areas. If for some reason all the teams in one brigade were—say in Belfast—committed or there was a shortage of leaders, perhaps because of rest and recreation leave, I tried to find replacements from my own staff. Only once in 14 months did I have to ask 3 Brigade to help out because of events in Belfast.

And then there was the Theatre Reserve.

When I first came across the phrase I was impressed. The

words somehow conjured up a veritable 'old guard' of bomb teams, steely-eyed veterans with nerveless hands emerging from camouflaged hangars in convoys of powerful armoured vehicles bristling with machine guns. They would have, of course, only the latest equipment. I saw them like a legion of Drakes listening for the Drum while slumbering in their DMS boots, or putting fine adjustments to impeccable Wheelbarrows, until the balloon really went up. And then they'd show 'em.

The truth was sadly different. The Theatre Reserve was to be improvised from anyone who happened to be around at the time. Its equipment consisted of a single Pig with a Wheelbarrow. The operator was to be 'anyone who was available'; the driver was a corporal clerk and the escort was anyone who could hold a rifle in the approved military manner It was a classic opportunity for the Duke of Wellington's famous phrase about his own troops: 'I don't know about the enemy but by God they frighten me!'

Naturally, when I learned of this I did not noise it abroad, but took consolation in the standing operational procedure which laid down that a period of time had to elapse before the Theatre Reserve could be called out. This would in theory, enable a balanced team to be provided.

Unfortunately, the grandiose terminology created a grandiose impression in other minds and on one difficult night when Belfast was having a particularly hard time the duty watchkeeper in the Ops room of 39 Brigade decided he would call out the Theatre Reserve. Alpha and Bravo were already deployed. Echo was somewhere to the north of the city, and he felt the need to keep Charlie and Delta teams under his hand. I gathered later that the officer responsible misunderstood the procedure – he seems to have thought that he had an unlimited number of teams at his disposal – and did not check first to make sure it had been brought up to a state of readiness. Just how ready it was I witnessed myself at the closest of quarters.

Inevitably I was making my way to Belfast when it happened. Soldiers with weapons waved down my car as I got to the bottom of Magheralave Road, leading from HQ Northern Ireland. Streets were being sealed off and there were all the signs of an incident. Having identified myself, I set off for the control point to find out what it was all about. On being told

43

there was a suspicious package on the steps of a showroom in the station approach, I took up a suitable vantage point and waited for the EOD team to arrive so that I could watch them do their stuff. It was a nice feeling. I had every confidence that soon the well-oiled machine for which I was responsible would purr into action. They would deal quietly and efficiently with the matter, the area would be cleared, the trains would begin to run again and, after an encouraging word with the chaps, perhaps a modest aside to any of the Headquarters officers who happened to be around ('All in a day's work....') I would be on my way.

Minutes passed. The silence deepened. I found myself peering up the road from which the team was most likely to arrive. Once or twice I thought I heard promising noises in the distance. Nothing. One or two of the soldiers forming the cordon, those to whom I had indentified myself, began casting the odd glance, or so I felt.

More minutes passed. Still nothing. More glances, questioning. After 15 minutes had gone by I made my way to the ICP and called up 39 Brigade Ops on my radio. Back came the solemn assurance from the Ops room: 'The team has been called out.'

Short of holding an inquest over the ether there was nothing I could do but accept this flat statement.

I went back to wait.

After a minute or two more, joy of joys. Through the balmy evening air came the confident, brassy tones of the EOD siren, growing louder by the second as the team streaked down the hill from Thiepval Barracks. Ah, my boys, my boys. The noise grew deafening and, with blue lights flashing bravely, a Pig swung round the corner and roared up to the incident control point. It came to a halt with a screech of brakes and stood there rocking on its springs like a faithful hound panting after the chase. All eyes, particularly those which had been following me so accusingly, were riveted to it as the door opened and out stepped ... a solitary short, tubby, bespectacled corporal. There was no-one else. There was no ATO, no escort, no signaller, no other vehicles, no nothing. *This* was the Theatre Reserve?

The onlookers – who were not to know that the new arrival was one of our excellent HQ clerks – stood transfixed waiting to see what miracles, singlehanded, he would perform. A one-man EOD team, this was something new. A couple of soldiers respectfully shuffled out of the way as this mysterious figure waddled purposefully towards the incident control point. The eyes behind the spectacles were giving nothing away and the pink face was inscrutable. Only when he was a few feet away did the set visage show signs of life. The men of the cordon craned foward to catch his words.

'Oh, hello, sir,' he said chattily. 'Are you here too?'

'Yes', I said, 'But where the hell is the operator?'

'Brigade HQ are trying to find one. Meanwhile I thought that I'd bring the vehicle down and meet him here.'

I was out on a limb. I couldn't just stand there passing the time of day with the driver. So I took over the operation. The glances began again.

It was not one of our most polished performances. The corporal, who had shown considerable guts and energy in deploying himself and the Pig, was not experienced in bomb clearance. But he was willing.

'Right, let's get Wheelbarow out, corporal.'

'Right, sir.'

He marched smartly to the rear of the Pig where there was a metallic rattling followed by a pause. The pink face turned towards me.

'It's locked, sir.'

'Get the key, then, and open the bloody thing.'

'Right, sir.'

The rotund figure marched purposefully to the cab and began rummaging ... and rummaging. The situation was beginning to get ridiculous.

'It's got to be here somwhere.'

'Yes, sir.'

'Try the tool box.'

'Right, sir.'

I joined in the search.

What in the old days were called 'lesser breeds' might have panicked or had hysterics in the middle of this nightmare. Not our RAOC corporal. Orders were orders. His colonel had said the key must be there somewhere and so there it must be. The man said search and so he searched, systematically and dog-gedly. The sublime confidence of the soldier that something would turn up was as impressive as it was maddening, but in the end something did turn up. Triumphantly we retired to the rear of the vehicle and opened the padlock. The doors swung open and we found ourselves staring into the dead, electronic eye of a Wheelbarrow.

Getting it out of the Pig was no problem. We put the ramps in position, fixed the short power cable and drove it onto the road. We stood looking at the thing for a moment then glanced at each other. The good corporal was waiting expectantly.

In theory, of course I knew all about the Wheelbarrow. I knew how it worked and what it was meant for and how it was used. I had even criticised operators for not using it when they might have done, but I had never used one myself in anger. Although I had had a go when I did part of the pre-ops course.

And, after all, I told myself, why should I? I was the boss, the specialist with the penknife, the man in the bomb suit, the fellow with the hook and line. None of your mechanical ops for me – all my jobs were handmade.

Be that as it may, I couldn't very well explain that to the assembled company. Besides, it looked easy. There would be nothing to it.

'Better connect the main cable, corporal.'

'Roger, sir.'

He did this with alacrity, put on his expectant face again and waited.

After a bit of trial and error a fuzzy picture appeared on the closed circuit TV screen.

'That doesn't look so hot.'

'Oh, I don't know, sir.'

The corporal sounded on the defensive as if he and the Wheelbarrow were chums.

'Well, here goes.'

The Wheelbarrow set off jerkily in the direction of the bomb and I manipulated the controls as if to the manner born. The picture was bad, however. This was because the programme to modify the Wheelbarrow with a better camera had not got as far as the Theatre Reserve equipment.

As dusk was marching apace, it did not help that the camera on this particular Wheelbarrow had poor low light capabilities. Halfway between me and the bomb the Wheelbarrow

46

suddenly stopped. I'd driven it into the rope that marked off a part of the road that was under repair.

'It's stuck, sir,' said the corporal. There was disappointment in his voice. 'Try bringing it back a bit.'

Well, I tried bringing it back, sending it this way and that plus a series of jerks, but succeeded only in making the mess worse.

At this point reinforcements arrived. The cheerful face of Sergeant M—, my sergeant and clerk, appeared at the rear of the vehicle. Sergeant M— and I had carried out a couple of operations on the border and he had done a tour in Bessbrook as an operator, so his presence was welcome. He was a highly competent, cheerful and experienced EOD operator.

He had been in the bath at home in his married quarter when he heard the Pig go out. After contacting 39 Brigade to find out what was going on he had dressed and driven down to Lisburn to see if he could help. Apparently he should have been called out originally, but someone forgot to send for him.

I explained the situation to him and told him to have a go at extricating Wheelbarrow. He tried but was no more successful than I.

The nightmare was getting worse.

We decided to walk over to the footbridge over the railway and see if we could see the object. Cars spoiled our view. There was now only one alternative – the clearance would have to be done manually. As I had started the task there was no question as to who would do it – we didn't exactly have a poll but it was obvious that I was the most popular candidate. First of all another problem had to be resolved. Both the disruptive devices were fitted to the boom of the Wheelbarrow and no spare was carried in the Pig. And we couldn't bring the Wheelbarrow back to the ICP. So, I had no choice but to go and get one from the Wheelbarrow.

I put on the bomb suit. With a torch in one hand and a tool kit in the other I set off into the gloaming and examined the Wheelbarrow. My plan was to remove the shot-firing device from the equipment, rewire it, take it to the side of the suspect package and fire it.

Half blind and crippled it may have been, but this particular Wheelbarrow displayed great affection for its possessions. It was extremely attached, it seems, to the disruption devices I

47

needed. Someone had clamped them very firmly to the forward tube and the Wheelbarrow refused to part with them. I heaved at it, and I heaped invective on it until eventually I achieved my object. Worn out, I returned to the ICP and rewired the contraption to the alternative firing cable.

The rest was easy. I went into the station approach, laid the disruption device beside the package, and came back to the ICP.

I called up Brigade HQ: 'Hello Zero this is 15 Zulu. Small explosion in my location in figures 2 minutes. Over.'

'Zero. Roger, out.'

The report of the explosion startled a few sleepy pigeons into the air but that was about all. The package was a hoax, stuffed with cotton wool.

After I examined it I searched the rest of the area, returned to the Pig, handed over the bomb suit and my bits and pieces to Sergeant M—, and declared the area clear.

M— went back to his quarter, I continued my journey to Belfast and the imperturbable corporal took the Pig back to HQ in solitary state. The Theatre Reserve was returning from action.

The following day I learned that someone had complained because our explosion had broken the plate-glass window of the shop.

'That's nothing,' I replied. 'The operation nearly broke my heart.'

The Baleful Border

Never at a loss for a cliché or the voices to overwork it, the Provisional IRA spokesmen lashed their tongues to a frenzy promising that 1976 would have 'a long, hot, summer'. They had suffered a number of setbacks along the southern border after the arrival of units of the Special Air Service Regiment and were anxious at least to give the impression that they had something up their sleeves.

From our point of view it was clear they would, according to usual practice, attempt a propaganda coup in the very area where they had lost the most face. South Armagh is one of the saddest areas in Northern Ireland because it has such a beautiful landscape. From Newry, which is on the border and has a much-bombed customs post, one can see the Mourne Mountains to the east. To the west lie the handsome slopes of Slieve Gullion. According to the seasons and the state of the broom and gorse and the heathers, the countryside is tinted with wholesome shades of yellow and purple. Small farms and cottages dot the valleys which are criss-crossed with streams and brooks ... gurgling through culverts which provide a natural hiding place for bombs.

Nowhere is the rural charm greater than west of Slieve Gullion where lies the little market town of Crossmaglen. The notorious market town of Crossmaglen lies in the heart of what the soldiers call bandit country – a farming area where the staple crops include shooting, bombing, murder and intimidation, not necessarily in that rotation. Furthermore they flourish regardless of season or weather.

For part of my tour the Security Forces base in and around the old RUC station was garrisoned by a company of 40 Royal Marine Commando, who took over during the climatically balmy days of August. To me, the base at Crossmaglen was not unlike my idea of a fort on the old North-West Frontier. Armed sentries peered from observation slits in lookout towers and movement inside was screened by stout walls and corrugated iron sheets. All movement in and out was done at speed and under cover of the weapons of the occupants.

In winter the place was an appalling sea of mud and the troops made their way between the huts by means of duckboards and plank walls. In summer the base was baking hot and dusty, though the troops always seemed to find the energy to play handball on a scorched strip of earth. The only person who didn't seem to mind the conditions was the ancient Oriental gentleman who ran a cigarette and sweets kiosk in one of the huts and, as far as I could gather, never went out.

Some of the troops amused themselves by taming the young jackdaws which fell off the RUC station roof and training them to sit on their shoulders or heads, but there was not a great deal about the place which one could describe as homely.

Because of the risk of ambush and the ease with which the IRA could set up illegal vehicle checkpoints on the country roads leading to 'XMG', all visitors went in by helicopter. So did supplies. Even the dustbins were dealt with by air.

Pumas and Wessexes flown by Royal Air Force pilots lifted patrols in and out of the base. They flew low over the countryside, dropping a section of men in one place and then putting down a second stick in another, picking up the the first after an interval, putting it down again and then repeating the operation. These tactics kept the opposition on the hop and made it impossible for them to be able to rely on any road being safe at a particular time.

For me, arrival at Crossmaglen SF base was always an interesting experience. A Scout or a Gazelle of the Army Air Corps would drop swiftly onto the Gaelic Athletic Association Football pitch opposite the back gate. You leapt out and ran towards the corrugated iron walls which opened as you reached them and were slammed shut when you were safely inside. By that time the 'chopper', the rotors of which never ceased, had whirled up and away.

The town itself contains a market square – not far from the base – more or less open at one side, and has a number of streets of stone-built houses which are fairly easy to obstruct. Usually a Saracen armoured personnel carrier is to be encountered on the approaches to the town. It is a difficult and dangerous area to patrol; narrow lanes with sharp bends lead into the countryside and in particular to the village of Cullyhanna, not nearly as well-known as Crossmaglen but equally steeped in villainy.

Attacks on the base at Crossmaglen continued throughout my tour – the first casualty I heard of was the paratrooper killed there by a radio-controlled device in a bicycle pannier – and there was no doubt that the operations mounted from the base that year got under the skin of the Provisional IRA.

All the more reason for them therefore to attempt a coup, something which would get them the attention they craved. They made their move at the end of August.

I was enjoying myself at the expense of a correspondent of *The Guardian* when the call came. The media were frequent visitors to HQ Northern Ireland – under the aegis of the Army Information Service personnel there – and this *Guardian* correspondent had invited me out to dinner. One of my several duties was to up-date the Press on the types of terrorist device being used and their efficiency. We had reached the coffee stage in a smart restaurant on the road to Belfast, and no doubt I was about to be subjected to the subtlest questioning techniques, when I was summoned to the phone. It was the duty officer from the Ops Room at Lisburn. Would I get down to Crossmaglen as fast as possible? Some sort of new weapon had been used in an attack on the place and my presence was required.

'Right. Lay on a chopper will you, I'll be with you shortly.'

The *Guardian* correspondent had followed me to the phone and was signalling vigorously.'

'Me too? Me too?'

You don't normally invite correspondents along to emergencies, but this one knew something was up and I didn't want the rest of the Press in the hotel to start getting the wrong ideas from a rumour. War or no war things are still done by the book in Northern Ireland, so I asked the duty officer to clear it with the Information Service for me to take a correspondent and we set off.

Coupe, who had been dozing peacefully in the Hillman, clamped our 'Kojak' style blue light onto the roof and we sped back to headquarters, sirens blaring. The main-gate check-point had been alerted and let us roar straight through to the helicopter pad on the sports field. A Gazelle was waiting with its rotors moving and, after telling Coupe to get some sleep and expect a call the following day, we took off.

It had been a clear day but the weather changed as he headed south. The strings of brilliant orange and blue sodium lights marking the main roads and the housing estates blurred under thin cloud and rain only a short time after we were airborne and by the time we reached Bessbrook, our first port of call, flying conditions were most unpleasant.

Bessbrook, a Victorian textile mill on the hills above Newry, is the headquarters location of the battalion responsible for Crossmaglen. Like Crossmaglen it is well defended and on the other side of the road is a metalled parking area which boasts of being the largest international heliport in Europe. One of my EOD teams was based at the mill permanently and when I reported to the CO of 40 Royal Marine Commando I was told that they had already flown into Crossmaglen.

All that the CO knew was that there had been a bombing attack on the base which was different from any previous mortar attack on Crossmaglen. No-one had been hurt and there had been only a small amount of damage but . . . the type of bomb involved seemed to be different from those used previously.

By this time the correspondent from *The Guardian* was greatly excited (though the CO of the Marines was less than happy at having a newspaper reporter land on him out of the skies) and became even more so when, soon afterwards, we set off from the heliport again.

Alas for the scoop. It was absolutely impossible for the Gazelle to get through the weather and over Slieve Gullion and the surrounding hills. There was nothing for it but to return to Bessbrook. From there I phoned Staff Sergeant B—, the ATO leading the team which had gone to the base, and obtained a description of the bomb. It did not correspond with anything we had found before. My conversation over for the time being, the indefatigable Coupe was called out from Lisburn to pick us up and, after dropping off my journalist friend, we returned to the barracks.

That night, while we were driving back through the darkness, three young men were dealing coolly with a number of deadly missiles. The garrison of Crossmaglen had been going about its usual business on the evening of 31 August when a loud report was followed by a series of shattering crashes so close together they sounded like a drawn-out single noise. When the dust had settled and the last splinter windmilled to the ground the garrison cautiously took stock of the situation. At four places they discovered the remains of some form of mortar bomb. That was the good news. The bad news was that five more were suspended in protective netting stretched above the buildings at the base.

The chain link nets had been introduced into Northern Ireland soon after it became clear that mortars were going to be an element in the campaign. Not unlike a form of wire mattress, the netting was stretched above the target buildings at a height determined by the estimated speed of action of the bomb's fuse. The type of fuse used in the PIRA bomb was similar to that used in aerial bombs during the war – there was a small impeller (a propeller working in reverse) on the nose of the projectile which screwed itself tighter as it flew through the air. In simple terms this pushed a striker into the body of the bomb and on hitting the ground a rimfire cap banged against it and sent a flash up a tube onto a simple, non-electric detonator.

The five bombs hanging in the netting above the huts at Crossmaglen were a testimony to the calculations made by Ministry of Defence scientists who had gauged its functioning perfectly.

A different form of expertise, however, was required by Staff Sergeant B— and his men after they flew into the base that night.

It was dark and the use of lights had to be kept to the minimum because snipers were known to operate in the area. The bombs were of an unknown variety and were tangled in the wire and its supports about 12 feet above the ground. Five of them had been located but there was no knowing if others were lying in the shadows.

The possibility of a second attack could not be discounted. Nevertheless, aware of the urgent need to identify the missiles and conscious that they were a menace to the whole base, Staff Sergeant B— and two NCOs climbed onto the threatened buildings and began to snip through the mesh holding the mortar bombs. His two helpers were a corporal ATO, serving as his No 2, and an acting sergeant who was the Weapons Intelligence NCO living with them at Bessbrook.

Below and in the observation towers, Commandos with rifles and light machine-guns at the ready waited for the first sign of the flash of an enemy weapon.

The snap of wire cutters broke the silence at irregular intervals.

Early the following morning I set off from Lisburn carrying descriptions of all known PIRA bombs, plus technical drawings of them. This time there was nothing to stop me flying in via Bessbrook and I did the usual final sprint into the base in fine style clutching the bomb data and plans like a rugger ball.

The staff-sergeant looked very tired by the time I flew in with the CO of 40 Commando. He had been working throughout the night on a task that was taxing both physically and mentally. After I had spoken to the base commander he led me to a corner, beside one of the huts.

'What do you think of these, Colonel?'

The staff-sergeant stood back rather proudly as if showing off his favourite dog's latest litter of pups.

There was nothing endearing about the sight that met my eyes. Four twisted bits of metal represented the bombs which had gone off. Alongside them lay five shiny black metal cylinders with a cluster of fins at one end and a stubby snout at the other. The bombs, subsequently dubbed PIRA Mk 8, were 39 inches long from nose to tail and weighed 15 lbs. A gadget had been introduced so that a collar with eight fins welded to it slipped down the body of the projectiles to give added stability.

The area inside the base had been searched already but we went over it once again without finding signs of any more bombs. I asked if anyone had been into the loft. It was realised that no-one had gone up there so I decided to take a look for myself. I called for a ladder and heaved myself through the hatch space.

There was more light than I expected and for a very good reason – bomb No 10 had made a hole in the roof and now hung by its tail from the wire perpendicular to the floor with its nose and sensitive fuse only six inches from the boards. It was the most evil-looking thing I had ever seen. I can best describe it by saying that it looked like a black snake ready to strike. It made me feel quite cold and I felt the back of my head start to tingle.

Calling to the staff-sergeant I told him to evacuate the building, an order complied with immediately. It then became a question of getting the bomb out of the roof space without jarring its nose. The impeller was well and truly screwed home and it was obvious that if it were given a sharp bang the inertia pellet which contained .22 rimfire cartridge would bounce on its creep spring and set the thing off. Somehow we had to free the tail without letting the delicate business end bang on the floor. Nor did I want the bomb to wobble too much or fall on its side . . . it was in too sensitive a condition.

If we could free the bomb from its entanglement we then had to ease it through the loft hatch, an aperture that was fairly narrow, and down the stairs into the open.

Without room to stand up, B— and I crouched under the dusty rafters and surveyed the scene. A number of water pipes were visible and the junction boxes of electric light cables. A water tank and a chimney stack plus a lot of cobwebs completed the picture. I also noticed some loose house bricks, left there I know not why, perhaps from some long forgotten repair job.

'Right Staff, I've got an idea.'

He looked at me doubtfully.

'I'll stick a brick under the end of the bomb so it can act a bit like a jack while I take the weight and you can cut the tail free.'

'And then . . .?'

He was very polite.

'And then I can lay it gently on the floor while you get back through the hatch and together we'll get it down the ladder and out.'

'Down the ladder and out?'

'That's right . . .'

I don't know what the doughty NCO was thinking at the time but in his place I'd probably have been saying to myself – 'I've managed to get five of these things out without accident so far; I hope to God he knows what he is doing.'

Regardless of what may have been in his mind, B— nodded in agreement.

'OK, Colonel, I'd better go and get the wire cutters.'

My position for this exercise was on the floor, sitting alongside the bomb. B— knelt above me. As I edged the brick under
the impeller he steadied the tail — swaying might be as
dangerous as dropping at this stage. Finally, when the vital
brick was in position I put my arms around the bomb and
cuddled it. Its black painted sides felt cold and clammy
against the side of my face.

'OK Staff, I've got it.'

'Right, Colonel! I'll start cutting when you give the word.'

'Just make sure you tell me what you're doing . . . I want to
know what's going on all the time.'

This was essential as I could not see a thing above me, being
eyeball to eyeball with the bomb.

Steadily my companion got to work, snipping and clipping,
the strands of wire parting with a twang and the weight in my
arms growing heavier as the tail was freed. Bits of grit on the
floorboards began to make themselves felt through my combat
trousers.

Throughout the staff-sergeant kept up a running commen-
tary.

'How long now, Staff?'

'About four more strands and then it's all yours, sir.'

'Thanks very much. Just take it easy. This bloody thing feels
like a ton. . . .'

And then, 'This is it, sir!'

I did not know whether to be relieved or terrified as the last
wire was severed. No miser ever clutched his hoard, no drown-
ing man his straw, no Lochinvar his bride, as I clung to that
bomb. It was mine, all mine, and God alone knew how glad I
was going to be to get rid of it.

As arranged, the staff-sergeant clambered around me and
down through the open hatch. Then he stood with open arms,
staring up, a study in concentration.

'Easy does it.'

One slip now and all our efforts would have been wasted.

'Got it. . . .'

B—'s hands took the nose and I steadied the tail as I came
out of the loft. Getting the bomb down two flights of narrow
stairs and into the open was difficult but nothing compared
with the affair in the roof space. What he and his men had gone
through cutting the bombs from the mesh in the dark I leave to

the imagination.

'Right, let's get rid of this little lot.'

But that was not as simple as it sounds. There were houses around – apart from the base buildings. And the bombs were big. In the end we took them down to the Gaelic football pitch. There was a concrete shelter there where we could take cover. They were sent up one by one and not a window in the place was broken. If anyone was upset about the large hole in the football pitch they didn't think it wise to tell me at the time. We took photographs and I took the bits back to Lisburn with me. I was not prepared to take a live bomb with me in a helicopter. We managed to work out the make-up of the bomb which was verified when some more were found in a cache. So much for the bombs themselves. What about the firing point?

From the strike pattern we had a good idea of the direction from which the projectiles had come. A patrol of 40 Marine Commando confirmed this soon afterwards, reporting a suspect flatbed lorry in a courtyard of the town square. It was about 100 yards from the base.

New problems presented themselves. It would have been simple for us to bowl over to the lorry and start poking about – but most unwise. Not only could the vehicle be booby-trapped, but devices could have been planted on the approaches to the lorry or in buildings near it.

It was clearly a case for a deliberate sweep and luckily we had a Royal Engineers search team on hand, flown in from Bessbrook where they were permanently stationed. We went with them in the Saracen to the courtyard. Because it was XMG we had to get a move on. Stand still for too long there and you could lie still for ever. For a time we became onlookers as the sappers, using detectors, probes and a variety of equipment, closed in on the lorry and checked the surrounding houses, reporting by radio all their moves and doubtful objects. When it was over we were able to move up to the lorry – a 3-tonner.

What we found was most interesting. Ten mortar barrels had been clamped into a purpose-built wooden frame and the whole assembly had been bolted to the floor of the lorry. Wires led from the bottom of each barrel to an electronic sequence switch. The circuits were activated by a standard PIRA timing and power unit so that the weapon could be armed remotely –

i.e. made ready at a distance – before firing also remotely.

The last terminal of the electronic sequence switch had been wired to a self-destruct mechanism placed above the diesel tank. This had worked all right, but the fuel had failed to catch fire and we were left with all the evidence.

Our final job was to dismantle the frame and remove the barrels. I handed one of the barrels over to the Marines and it is now in their museum at Plymouth, together with a replica bomb.

The mortar is a basic weapon with a short range but a high trajectory. It was used extensively in the 18th century as a siege weapon and it operated with some success at the siege of Sevastopol during the Crimean War (1854–5). The development of field operations led to a decline in its popularity, however, and by the time World War I began the British Army had none at all. As the war progressed, the need for mortars became urgent, especially as the Germans had retained the weapon in their arsenal primarily for attacking the Belgian and French fortresses. For a time the British were reduced to using ancient pieces of French equipment which bore the cypher of Louis-Philippe and had seen service during the Crimean War, but gradually an excellent weapon called the Stokes Gun was produced. It was a simple tube into which one slipped a bomb with a cartridge fitted to its base. This exploded on hitting a striker at the bottom of the tube and, with luck, the bomb was propelled into the enemy trench.

Mortars have remained a potent weapon in the hands of armies the world over since the First World War and in the last war they were reported to have caused more Allied casualties in the campaign in North-West Europe than any other weapon. The German Nebelwerfer, a multi-barrelled piece of equipment which could be towed easily, was particularly effective and it is quite possible that the Provisional IRA had this in mind at one stage in the development of the campaign.

Terrorist mortars were first reported in 1972 when a couple were unearthed by searching troops. Nearly 300 were seized the following year along with 100 bombs. Only a handful of mortars but a fair amount of ammunition was found in the two subsequent years.

The first home-made bombs were crude, unreliable affairs but, as the campaign continued, the weapons and the ammunition improved. By 1976 they were fairly sophisticated.

Attacks on a variety of targets occurred during the year – two in February, four in March, seven in May, three in June and one in July. Of these, the most significant took place on 6 March when a salvo of ten bombs was fired at Aldergrove Civil Airport, fortunately without causing serious casualties or damage. The mortars had been mounted in sand on the back of a vehicle but unlike previous weapons these had been fired electrically . . . like the German Nebelwerfer I have described above.

The bomb commonly in use at the time was what the Security Forces termed the PIRA Mk 6. Its overall length including the tail spigot and fin was 16 inches. It contained about 5 lbs of a home-made explosive nicknamed ANAL and was propelled by an incendiary mixture of sugar chlorate which could be adjusted according to the range required.

There was no doubt in my mind that the terrorists using the mortars had been given a good grounding in the use of the weapons and I remember a correspondent from *The Times* saying at a briefing at Lisburn that he had reliable evidence that PIRA active service units had been given instructions on ranges in North Africa. Occasionally manuals fell into our hands, one of which gave the angle of the barrel of the mortar (or mortars) with the charge required to achieve a specific range, i.e. one bag for 300 yards, two for 600 and three for 800 yards on an angle of 60°.

In the case of the attack I have described at Crossmaglen the men using the multi-barrelled mortar did not even have to bother working things out on the spot. It seemed quite clear to me that range and accuracy firings had been carried out on several occasions before the vehicle was moved into Crossmaglen. As the noise of training exercises could not have passed unnoticed or unreported north of the border you can draw your own conclusions where they did take place.

It is likely, also, that the operation was rehearsed covertly in Crossmaglen a day or two before the actual attack took place. I can see no other way in which the lorry could have been placed in the precise firing position in the courtyard. They used the chimney stack and radio aerial in the RUC Station as the aiming post.

It was essential for the Provisional IRA to take these precautions to prevent such attacks being counter-productive, in other words to avoid blowing to shreds sympathisers or potential supporters. For example, during an attack later in 1977 on the Security Forces base at Fort George, on the West side of the Foyle at Londonderry, their base plate positions were badly prepared. One bomb fell short of the base, two went into the predominantly Republican area on the far side of the road and one bomb went backwards.

There was also a near-tragedy when five bombs were fired at the Fort Monagh security base in Belfast in January 1977. Though two of the bombs failed to explode the salvo descended in the area of a school which was being used by a youth club at the time.

Two further attacks with five-barrelled mortars using Mark 8 bombs took place in the Newry area during my tour. On these occasions the tubes were in a pre-fabricated metal firing frame; from tool marks and other evidence it was deduced that all had been made in the same place.

The bombs had been altered slightly since their use in August 1976 – a wooden collar had been fitted across the shoulders of the projectiles to improve stability and the impeller fuse had been adjusted to reduce the incidence of blinds which had given Staff Sergeant B— such a massive problem.

This alteration in the setting of the impeller fuse worked up to a point but created another problem for the terrorists – premature explosions. Despite this ten bombs were fired at the base in Crossmaglen yet again in October 1976 and this time they all went off, and at the right time. Five members of the Security Forces were injured and one man lost his hand. Once more I was on my way to XMG.

The evidence showed this to be a much larger and simpler bomb than anything encountered before. Basically it was a 7 lb CO_2 gas cylinder with a plug welded to the base which held a tail unit with four fins.

Filled with ten to twelve pounds of home-made explosive, CO-OP, it weighed up to 35 lbs and had a range of about 100 metres. CO-OP is particularly sensitive and the firers must have been very conscious of this. Furthermore a bomb of that weight needs a very large charge to get it out of the barrel, let alone provide the wherewithal for range and accuracy. So the attackers discreetly used a remote control device to fire the

60

weapon ... after once again having positioned the flatbed lorry at the required distance.

The new bomb, designated by us the PIRA Mk 9, was fired from a nitrogen pressure tube of the kind used in factories making stout in the Republic. The CO_2 gas cylinder which provided the bomb casing is used, among other things, for the dispensation of stout and lager beers.

I am pleased to say that both beverages were among those consumed at the celebrations which followed the official recognition some time later of the gallantry of the men who cut out the unexploded bombs from the wire mesh at Crossmaglen. The staff-sergeant's share of the honours was a George Medal. This action was only one of several for which he received a recommendation for gallantry.

CHAPTER SIX

The Donegall Pass Dragon

It was a cold, dark, uncomfortable morning. Great drops of condensation dripped from the armour plate of the parked Pig. A little group of us huddled at the back in the shelter provided by the open doors, our eyes fixed on the silent film running on the small television screen. The sergeant's face was set in concentration as he manipulated the switches of the Wheelbarrow and its camera.

The little remote-controlled vehicle set out obediently enough up the deserted street, its shotgun pointing into the unknown, its cable snaking smoothly behind.

There was something gallant about the way in which the spindly contraption headed relentlessly towards the monster looming darkly through the gloom.

'St George and the bleedin' dragon,' muttered one of the watching soldiers.

The dragon reference was certainly right. This one was definitely the sort that could breathe fire.

It had begun the day as an ordinary petrol tanker. About ten o'clock that morning its driver had pulled in to a garage in the Whiterock Road, West Belfast, and begun to discharge his cargo.

As he walked into the office with his delivery note, however, the door was shut behind him and a gun thrust at his head by a man with a mask over his face.

'Sit down, you,' said the man with the gun. 'Over there facing the wall.'

Belfast being that sort of place the driver sat down, faced the wall, and started to worry. Other members of the filling station staff shared his anxiety as more masked men bustled about the place.

After about 20 speculative minutes the man with the gun

said, 'Right. Get back in the cab and drive that bloody thing to the RUC station in Donegall Pass.'

The time-honoured warning about not trying any funny business was given but the driver's assurances must have lacked conviction because he was provided with an escort. Ahead of him went a Black Taxi – a vehicle resembling a London cab but which operates unlicensed in certain areas carrying as many passengers as it can pick up for a few shillings each. In the rear came a battered Cortina.

Donegall Pass is a broad street running off a busy junction where Great Victoria Street meets the Lisburn Road and Dublin Road. Like all RUC buildings in Northern Ireland, the police station had armed guards, wire netting and concrete to protect it and was a hard target for the normal terrorist operation. It had been made almost impossible to get near enough to do damage with an ordinary bomb, which is no doubt what prompted the Provisional IRA to use a tanker as a super-incendiary.

It was very foggy that morning and with the traffic as heavy as usual, the gunmen had little difficulty in halting the vehicle opposite their target. One of them jumped from the lead car and ordered the driver out before driving off. The man immediately ran into the building and broke the news: he'd been hi-jacked – not a sensational occurrence in the context of the times – and he thought there was a bomb in the tanker. It was 10.36 when he arrived. Two minutes later there wasn't a soul in the building.

My mind was actually on a personal shopping errand when the thin dribble of static on the car radio became a full flood containing my call sign. I told Coupe to pull into the kerb, picked up the mike and pressed the switch.

'Hello one five Zulu. This is one five. Over.'

'One five Zulu – Send. Over.'

'One five: can you give your location? Over.'

I explained that we had just come off the motorway and were in Donegall Road heading in their direction.

'One five: Roger.... We have a problem.... Suggest you postpone your visit and carry on in the direction you are travelling in ... you should find something of interest quite close to your location very soon. Over.'

'One five Zulu. Wilco ... out.'

'Something of interest' had an ominous ring about it.

So much for our shopping expedition. It had been my firm intention to buy a Valentine card for my wife after visiting the EOD sections which had moved from the famous Albert Street Mill to a more salubrious residence in the old Post Office.

'Whatever happens don't let me forget, Coupe,' I told him. 'I don't want to go back without a Valentine card.'

'Right you are, Colonel.'

'OK! You heard what the man said. Let's drive.'

The first sign of 'something interesting' was a cordon of policemen and soldiers across the road near the Sandy Row area. Gunners of 49 Field Regiment were squatting in doorways nursing their rifles while others, with men of the RUC, were diverting traffic.

We were in civvies and after a brief explanation as to who and what we were, we were waved through. Finally I learned from two officers about the tanker outside the police station. Coupe moved our vehicle smartly into cover behind the buildings on the corner of Dublin Road.

'It's this damn fog,' said the RUC inspector in charge at the scene of the incident. 'You can hardly see the tanker from here.'

It was, indeed, hard to make out anything more than a blur at 100 yards, which was the range from which we were viewing the menace. Then the sound of a siren split the air.

'That'll be your boys,' said the inspector. 'I think they got stuck in the traffic.'

Though they were only a mile and a half away in Royal Avenue, Bravo bomb team from the Central Post Office had been forced to crawl through the traffic held up because of the bomb threat. When eventually they arrived, at 10.55, they were on the opposite side of the tanker to myself.

I weighed up the position and decided that Sergeant A—, who commanded the team, should bring his men and vehicles to my end of the street where they would have more protection from blast if the tanker blew up. I called him up on my radio. 'Hello one five Bravo, this is Zulu. Move to this end of Donegall Pass. Over.'

'One five Bravo – Roger. Over.'

Almost immediately there was a very loud bang, and everyone ducked. I grabbed the radio again and called the section commander.

'One five Zulu, what the hell is going on? Did you shoot at something? Over.'

'One five Bravo Negative. Over.'

'One five Zulu are you being fired at? Over.'

'One five Bravo Negative. Over.'

'One five Zulu, OK get here as quickly as you can. Out.'

No-one else reported either firing or being fired on so for the moment the noise remained a mystery.

By this time we were beginning to gather quite a number of people directly concerned with the fate of the tanker, including the head of the Northern Ireland Fire Service, from Lisburn, and the local fire brigade chief. The City Surveyor and a manager from Burmah Oil, who owned the tanker, were also on their way. Through the police I asked for a number of lorry-loads of sand to be made available.

The City Surveyor was needed to advise on what problems might be created if, for some reason, the petrol got into the sewers and the sand was a good standby to build barriers if it was found necessary to stem a flow down a road.

While the representatives of the various agencies involved were on their way, I was able to question the tanker driver. Fortunately he was a sensible and observant man who had kept his nerve during what must have been a nerve-wracking drive.

The tanker, he revealed, carried 3,500 gallons of petrol, and its tank was divided into seven compartments. The three nearest the driver's cab were empty; the others were full. The driver had been unable to see what, if anything, the hi-jackers had placed in or on his vehicle but, when the gunman ordered him out at the police station he had seen him pull a string which seemed to lead to the top of the tanker.

I questioned him closely on the contents of his cab, but he was adamant that the hi-jackers had left nothing in it.

'There's only my log sheet and my tool box in there, sir,' he said. 'You can be certain sure of that.'

As far as he could see, all the caps on top of the tanker had been secure when he left the filling station.

Having exhausted my questions I thanked the driver – it pays to be polite – and went back to Bravo team. From what I had been told I came to the conclusion that the device was in one of the empty compartments in the front of the tanker where the petrol vapour would increase the violence of the explosion. I was also fairly confident – though not 'certain sure' – that the timing, arming and power unit was somewhere up on top of the tank, possibly with the bomb. We really needed a good look at the vehicle to confirm these suppositions.

By this time we had deployed the Bravo team across the entrance to Donegall Pass with its Saracen and two Pigs in herring-bone formation so that we could shelter behind them while mounting our attack. More than an hour had now passed and we seemed to be progressing at a snail's pace. It was at this point that we sent Wheelbarrow into action.

The little machine and its operator did their best. But though it jerked this way and that on its tracks so that the closed circuit television camera could examine as much as possible, we would see nothing of interest on the screen in the back of the Pig. The fog was too thick, the shadows under the vehicle too dark. It was like watching a very bad silent documentary. Wheelbarrow was recalled but only temporarily.

Included in our armoury at that time was another remotely controlled device, an armoured fork-lift truck we called Eager Beaver, which could travel across country. It was the only one in the Province and I had stationed it in Belfast. I decided it was high time that we sent for it. Not without difficulty we mounted our faithful Wheelbarrow on the forks of the Eager Beaver in the hope that the added height would enable us to get a look at the top of the vehicle and, even better, attack it remotely. This ungainly contraption went rocking through the fog but after a few yards changed direction. It was stopped and pulled round. But no sooner had it gone a few yards more than it headed off again on a different course.

'I think it's trying to tell us something, Colonel,' said Sergeant A—. 'The ruddy thing doesn't want to go anywhere near the tanker. It doesn't like bombs.'

Strangely we were able to retrieve it without difficulty, but we were still no nearer adding to our knowledge.

There seemed to be only one avenue left. We would have to go into a building overlooking the tanker though this was against the rules. On the strength of the fact that as CATO Northern Ireland wrote the rules there was no-one better to break them I decided to use the threatened police station itself as an observation platform. Led by two stalwart figures from the RUC, the EOD sergeant and I entered the rear of the building and tramped through silent offices and along empty corridors and up deserted stairs. It was a flat roof and we were four or five storeys up.

Peering over a parapet blackened with the grime of years and splashed with pigeon droppings, we took stock of the situation. My skill with the monocular telescope (which is a standard part of an EOD man's kit) being, for some reason, not entirely satisfactory, I had to get A— to spot for me. By this time the visibility had improved enough for him to be able to see the top of the tanker. All the seven filling caps were closed, he reported, and 'there's a box-like thing near the front one'.

Unable to discern anything else of interest we returned in procession back through the police station to the control point.

We now had reason to believe that –
a. the device was in the front empty tank,
b. that the filler cap had been lowered after it had been inserted,
c. that the arming, timing and power unit was remote from the device – i.e. in the 'box-like thing'.

Deduction: if I could remove the TPU it ought to be safe to pull out the bomb by hand.

This assumed, of course, that there was not a secondary arming circuit inside the device. Having convinced myself, reluctantly, that this was unlikely on the facts as they stood, there remained only one thing to do.

'Sergeant ... I'll borrow your bomb suit if I may.'

'Be my guest, Colonel.'

All the bomb suits I have encountered have the same smell – a mixture of sweat, dirt and explosives, of which the predominant smell is sweat. Without wishing to cast any reflection on Sergeant A—'s personal hygiene, I will say merely that his was no different from the rest except for the fact that the odour of explosives was more pronounced than usual. If anyone wonders why I bothered to try to cocoon myself against the inferno which would result if thousands of gallons of petrol went up, I can say only that, at the time, any protection seemed better than none. Besides, the device might go up when I was some distance away.

It takes a few minutes to put on a bomb suit but on that occasion it seemed like seconds. The bracing up of the trousers, which go on first, and the careful adjustment to make sure they don't get twisted and cause discomfort – which can interrupt concentration – was over in a flash.

The front of the jacket with the high collar and pouches and the separate back section were on almost before I realised it. Bingo – the abdomen and chest armour was in position and one of the team was standing by with the helmet.

'No thanks,' I said, trying to sound cheerful. 'I've got my own.'

The truth is that because of chronic arthritis in my upper spine I couldn't hold up my head in the heavy regulation helmet that goes with the suit so I carried a lighter version in the back of the Hunter along with other bits and pieces. It was proof against a well-thrown half-brick, something I knew from experience, but that was about all. Unfortunately it meant I couldn't use the self-contained radio in the regulation helmet. But I didn't feel like talking to anyone anyway.

Bravo team regarded my helmet with thinly disguised contempt as it was produced.

I had already handed Coupe my civilian jacket and my pistol before I started to don my suit of armour. Now, like a squire seeing off a knight of old, he handed me my weapons ... a hook and a line. The rest of my tools were in the pouches.

It was time for a final check. The evacuation of the buildings ... complete. The cordon in place. The fire engines ... ready and waiting. Cover against snipers ... OK.

I could think of no more excuses to delay what had to come ... there didn't seem to be anything to say so I didn't say it. I had walked towards bombs before but never to one in a petrol tanker, though I had seen the results of one which had exploded. Better set off. A member of the team said, 'Watch your step, sir.'

The 100-yard walk seemed quite short, despite the 60 lbs hanging on my torso. Automatically my eyes scanned the vehicle which shed its fuzzy edges as I got nearer – the fog was thinning. All the time I kept on asking myself what was the best way to get onto the tanker without disturbing anything. It looked sullen and dangerous and challenging, its metal sides streaked with grime, the great tyres motionless and sulky with the dirt from Ulster's troubled roads. When I got close enough to be able to touch it I put my hook and line on the ground and started looking under the wheel arches, the belly, the engine and round the uninviting sides of the machine. I saw no more than Wheelbarrow had shown us.

The cab, which I searched next, was as the driver had said. Only his log book and tool box were visible. There was no sign of the string the hi-jacker had pulled.

I retrieved my own line and the essential hook and plodded to the iron ladder leading to the top of the tanker.

It took me some time to get one foot on the first rung for the simple reason I could not see easily, what with the high collar of the bomb suit and the other restrictions to one's movements. Having got one foot on the first rung it was even more difficult to find the next step, holding my hook and line and heaving up the extra weight of protective clothing and armour. By the time I had made it I was beginning to add pints of my own perspiration to the now definitely anti-social bomb suit.

The cap on the first tank had not been properly closed and was slightly ajar. By the side of it, just as A— had described it, stood a rectangular plywood box, readily identifiable as a piece of Provisional IRA equipment.

Acutely aware that the fog had made the steel slippery I took great care as I bent down and slipped the hook into the box. Gingerly but ponderously I retreated down the ladder making sure as I did so that my line was not going to get fouled. Once on the ground I retired for a distance of some 100 yards to the rear of the vehicle. Some two and a half hours had now elapsed since the bomb had been delivered and this was the first direct action I was able to take. What would happen when I pulled the line I could not know for certain, but according to my calculations . . . .

I pulled the line sharply and felt something give.

The ascent of the ladder was no easier the second time but at least I had the satisfaction of finding, once I reached the top, that I had completely dislodged the box from the side of the closing cap. A blackened detonator on two wires hung from one end of it. That explained the 'shot' we had heard in the fog. In retrospect, it appeared that the terrorist had dislodged the detonator when he pulled the string which removed the wooden dowel pin holding apart the jaws of the arming switch.

Clearly it had been intended to send up the whole show at

the time we heard the crack.

Rather encouraged by events so far I opened the filling cap very gently and looked inside. Everything was suddenly not in order. Apart from a hole I could see a pipe which according to my knowledge of tankers should not have been there. After an incident on the border on Christmas Eve I had taken the trouble to inspect petrol tankers at the Shell depot in Belfast. Now I needed to know if Burmah tankers were any different. There was nothing for it but to complete once again the descent of what was beginning to seem like Everest ... steadily, one foot after the other, feel for the ground, and then plod, plod, plod, back to the incident control point.

Expectant faces greeted me, and there were one or two anxious looks. For all they knew I might be bringing news of imminent disaster.

'Where's the tanker driver?'

He was produced but denied all knowledge of the pipe I described in the tank.

'Are you sure?'

'Certain sure!'

I drew a sketch in my notebook. The driver studied it thoughtfully.

'No, sir, that's one pipe too many.' He brightened. 'Maybe it's the bomb, sir.'

'Thank you. . . .'

This time the journey back did seem a long way, the height of the ladder even greater.

The pipe, I observed on close scrutiny, was held to the filling cap by a length of fishing line. Kneeling over the hole, I held it in one hand while using the other to fiddle with the fishing line which was wrapped around the hinge. I was there some time, and my trousers became soaked with condensation, but eventually the line was untied. From inside the tank I extracted a two-foot length of grey plastic pipe about three inches in diameter. The ends were sealed with black adhesive tape.

Clutching it tenderly I brought it down from my perch and tramped 50 yards from the tanker where I had laid it in the gutter. Extracting two disruptive charges from my pouches I placed one at each end of the device and from a safe distance, behind cover, fired them. The pipe split open.

There was still more to do but by now I was confident that

the danger was over. Moving back to the EOD vehicles I got out of the bomb suit and got the Burmah Oil manager to let me have the key to the filler caps so I could check them out. There was nothing in them.

At five minutes to two I was able to declare the area clear.

The remains of the bomb when handed over to the SOCO – the RUC Scenes of Crime Officer – included a sample of the 4 lbs of CO-OP explosive which had filled the pipe; a short-delay detonator and a complete IRA timing and power unit containing two 1½-volt batteries and a one-hour delay timer.

It was suggested later that I should have claimed salvage for handing back intact one petrol tanker plus its contents to the oil company. I pointed out that I wouldn't have liked to meet the bill every time things turned out differently.

As it was, Burmah sent a cheque for £25 to the RAOC Aid Society, which helps the needy, and I and the other members of the team each received a company tie bearing the emblem of the Chinthe, a mythical lion which guards inhabitants of temples and important buildings in Burma.

Most pleasing of all was the flood of letters of thanks we received from the people of Belfast.

Later I heard that examination of marks on the recovered detonator revealed that it had been made in Great Britain and had been sent to Irish Industrial Explosives in the Republic of Ireland for resale. The explosives also bore charcteristics which indicated they had been brought up from south of the border.

The Donegall Pass tanker, though I could not know it at the time, was part of a specific campaign by the Provisional IRA which spanned the winter of 1976 and the spring of 1977.

The tanker 'season' opened at Newry where the terrorists parked a hi-jacked vehicle complete with bomb outside the customs house on the A1 near the border. Before the EOD team could get to it (luckily for them) it went off with spectacular results. Apart from damaging the customs house, which was well back from the road, it set off a fire ball that rolled down the hill and destroyed five cars *en route*. The petrol container was blown wide apart and the heavy cab was hurled into the middle of a field on the other side of the road. Despite the force of the explosion and the severe damage, no one was hurt. Nevertheless, after inspecting the wreckage and the blackened gutted cars, it was quite clear to me that under different circumstances a tanker 'bomb' could cause heavy loss of life. I fully expected the Provisionals to introduce another into their programme and though they moved in a roundabout way, they did not disappoint me.

What was designated Operation Cupar began in the first days of December when a car was discovered 150 yards from the H21 border crossing* with debris scattered on the road and in the field alongside. At first sight it had been the target of some sort of explosion, and a fairly large one at that.

A careful study of the wreck and the surrounding area revealed that to the rear of what was left of the vehicle were two undamaged milk churns, one standing up and one on its side. Beyond the grass verge was a thick and high hedge with broken ground on the far side. Telegraph poles lined the road, and about 100 yards away was a house with a drive leading into it. The whole thing smelled very strongly of a trap.

Obviously the milk churns might be full of explosives. The hedge and broken ground was ideal for an ambush party – don't forget the border was just a sprint away. Telegraph poles, along the wire fences, provide excellent concealment for the aerials that go with radio-controlled bombs. And the house with the drive was a natural place for the Security Forces to set up an incident control point which the terrorists could have zeroed in on, or prepared as a target in some way or another.

The weapons intelligence officer from 3 Brigade headquarters carried out a close covert scrutiny of the area and reported that there were no signs of bits of body near the vehicle or signs of anyone having been injured and so it was decided to leave the wreck where it was for the time being. Left alone it could do no harm and so it remained in splendid isolation ... until December 22.

The terrorists, who had been watching, undoubtedly waiting for something to happen, got tired and decided to speed things up – perhaps so they could have Christmas off. A tanker was hi-jacked at a petrol station on the border and driven to within 30 yards of the blown-up car. There it was manoeuvred across the road so the rear wheels were on one verge and the front wheels on the other. The road, which had

_____
* All border crossings are numbered on military maps.

been clear on one side of the wrecked car, was now completely blocked. In addition, the tanker was close to a school, a farm and a store. The area had to be cleared rapidly and the 1st Royal Highland Fusiliers, who were the battalion responsible for it at the time, decided it should be done on Christmas Eve. After all, the Scots prefer to celebrate at Hogmanay.

The problems raised by the vehicles were complex and varied. The Provisionals had deliberately fixed things to attract us into the area so we could take it as read that they had prepared something unpleasant for us. The question was where they had hidden their surprise package – the churns, the side of the road, the drive leading to the house or the petrol tanker. Was it radio-controlled, remote-controlled, a pressure plate mine or what? We did not know even how many there might be. In the circumstances, there could easily be more than one.

After weighing up the situation I told Warrant Officer L—, ATO at Bessbrook, that I would come down and do the job. While he and Sgt Walker, the Royal Engineers search adviser, got on with the planning at their end I would go along to the petrol farm depot in Belfast and find out what was normal and what was not normal in a petrol tanker.

The morning in the depot proved interesting and, as it turned out, invaluable. Borrowing a key for opening the lids of the hatches, I clambered all over the vehicles staring into their innards and exploring their cabs and fitments. The firm gave me every assistance and a lot of information before I left.

Early on Christmas Eve Coupe and I set off to drive to Bessbrook Mill where the ATO was waiting to tell me about his arrangements. After receiving his report, I went off to see the commanding officer of the Fusiliers, Lieutenant-Colonel Bryn Campbell, and got his briefing. He told me that his C Company were running the show and that four patrols of Jocks had been operating in the area the night before and three more, plus an incident control point, had flown in at first light. A cordon had been provided to prevent movement in and out by road and cover in the form of an airborne reaction force was also being provided.

The area had been flown over by air recce and nothing untoward could be seen on the air photos.

No-one could have asked for more and soon afterwards the

Jock sentry was opening the gate to let us cross the road to the heliport where our machine was waiting, its engines running. With us went the RUC liaison officer who was our link with the Gardai with whom we were co-operating. In no time we were looking down on the tall chimney of Bessbrook Mill and the parked vehicles. Then it was down to some fast low flying in a Puma to a spot ten yards from the border crossing.

The device we used for detecting the presence of radio-controlled bombs was working by the time we arrived and Sergeant Walker was ready with the Sappers to clear the area up to and around the tanker.

I got them to pay particular attention to the verges and the ground in the immediate vicinity of the tanker and they did a thorough job. An hour and a half later they reported their sweep complete. It was then up to me to go forward and check the tanker.

The fifth wheel, the cab, the steps leading to the top of the petrol tank and the exterior generally revealed nothing – I had been very suspicious of the fifth wheel. It was possible, there-fore, that a bomb might have been placed inside one of the petrol tanks and we decided to go ahead with a pre-arranged plan to open them remotely.

The high bank which had made the spot so attractive to the terrorists actually came to our aid.

One by one I was able to unlock each lid – but refrained from lifting them by hand. In turn I attached a hook and by means of a snatch block device which L— and I operated from the bank (at a distance) we were able to pull the lids open. In-vestigation showed the tanks were full of petrol but I could see nothing untoward. As far as I was concerned the tanker was clear.

Ever hopeful, we had brought spare batteries with us so we were able to start the engine. With a member of the EOD team who had an HGV licence, I took my place in the cab while he manoeuvred the vehicle off the verges and backed it up the road to a police station where it was handed back to its owners. The reason I sat in the cab was to show the driver and others I was completely satisfied that both the tanker and the grass verges were clear. So far, so good

The wrecked car now needed to be dealt with and once more the Royal Engineers search team got to work. We had to move our incident control point to a more suitable location, about 100 yards on the other side of the vehicle.

Time was beginning to press by now because of the short day. The light was already beginning to go slightly.

Once the sappers had done, and I was satisfied that there were no indications of a radio-controlled device, I moved in and put a hook and line on the standing milk churn. After pulling it over from a distance without anything happening we could be sure there was no anti-tilt switch in it and no pressure plate underneath. Together, WO2 L— and I walked back to the churns and taped a primer and detonator to the bottom of each. Sudden sharp pressure applied to the base of a churn should make it act like a mortar, ejecting whatever is inside. But on this occasion, after we had fired from a distance of about 100 yards, where we had established the new ICP, something wasn't quite right. A certain amount of what was evidently explosive had been thrown out, but not enough to satisfy us. I went back alone this time and saw that one of the churns still had something in the bottom. I was not prepared to put my hand inside to see what it was.

I called L— forward to help and we attached hooks to the base of each churn. With the invaluable snatch block we led the line over a tree on the verge and retired to the ICP. With a series of short sharp tugs we then banged the churns together and shook out the rest of their contents.

'Not very nice,' said L— when we returned to study the results of our handiwork. 'I don't think they like us.'

'You could be right.'

The plot was simple but effective. From the beginning the wrecked car had been the bait to kill an ATO. The tanker had been merely a ploy to bring us into the area.

The upright churn had been filled with explosive and had a length of Cordtex fuse running through the middle of it – nothing else. It was harmless unless a detonator was introduced. The horizontal churn also had Cordtex leading through the explosive but in this case it went into a timing and power unit. An unsuspecting, or overconfident operator, working against the clock because of the short day, might be duped into thinking that having cleared the one churn without trouble, the other would be the same. But had the cord in churn number two been pulled it would have fired the detonator of the timing and power unit which was fixed firmly to the bottom by a wooden slat.

With proof of the horribly sophisticated thinking of the enemy before our eyes, we were meticulous about our clearance of the wreck, pulling open the shattered doors and the boot remotely. Once everything had been thoroughly checked we rolled the remains into the hedge.

All that was left to be done was to secure the churns, the timing and power unit and samples of the explosive – it was ANFO with a CO-OP booster in the booby-trapped churn – for the forensic science experts. What was left of the contents of the bombs we moved into a heap by the side of the road and blew up so that it could not be used again.

'Not a bad day's work, Colonel,' said L— as we climbed out of our chopper at Bessbrook in the dusk.

'Too right, it's not.'

We had cleared a tanker, neutralised a bomb and identified a booby trap . . . and even made a profit on the deal. One of the locals, understandably pleased at not having his property or relatives incinerated by the tanker or blasted by the bomb, had smuggled us a bottle of Irish whiskey.

All of us, the EOD, the RE search team, and myself had had a nip and wished ourselves a Merry Christmas before we parted.

Well on the way back to Lisburn with Coupe, the radio crackled into life . . . the Bessbrook team were being called out to a suspect device in Newry.

Not a bad day's work.

CHAPTER SEVEN

Choose Your Weapons . . .

Though it is far from a sporting campaign, the analogies persist. The phrase 'own goal' – to describe a terrorist who blows himself up with his own bomb – has become almost standard military usage. In the same vein, I found myself considering the equipment used by the ATOs in golfing terms: it was vital to ensure that operators had sufficient 'clubs' to get round the course, and had the facility to deal with the various hazards, while at the same time making certain that there were not so many options in the bag that selection became confusing.

In this respect one had to bear in mind that the equipment was designed for use by the average ATO and some were more versatile than others. Furthermore the 'course' varied just as they do from club to club throughout the United Kingdom. Some men could get by on two clubs, so to speak; most had a half set; others used a full set and a few – rare birds these – had kits made specially for them. One 'club', however, was common to all – Wheelbarrow, the remotely-controlled tracked robot which carried a camera for closed-circuit TV and a boom, or extending arm, which could hold a shotgun, a spotlight or other devices.

Wheelbarrow was originally designed by a retired lieutenant-colonel working at a government research establishment. It began life looking somewhat like a pram, I believe, but could hardly carry such a name. Much of the praise for the development of what is one of the finest anti-bomb robot vehicles in the world must rest on the broad shoulders of Mr P—, the only civilian, as far as I am aware, who has the privilege of wearing the 'Felix' tie. This is restricted to ATOs who have successfully dealt with a live bomb – an ATO who had dealt with hoax bombs only during his tour is

77

not entitled to the tie.

'Mr Wheelbarrow' earned his tie by insisting that, 'You can't be satisfied a device will work unless you have tried it yourself.' According to the regulations, he was supposed to put in a formal application whenever he wished to visit Northern Ireland which I had to forward to HQNI for approval. I cannot, however, ever remember seeing such an application and he seemed to come and go at will, making his arrangements directly with the OCs of 321 EOD Unit. Because they never knew when he might turn up, the Belfast section always had a bed ready for him. This was so well used that on one occasion I received a complaint from his superiors saying that, appreciative though they were, he really should spend more time at his desk.

Just before I left Northern Ireland P— had almost perfected a method of controlling Wheelbarrow by radio, something which all operators had been anxious to see introduced. One of the limitations of the machine was that it moved at the end of a multi-core cable supposedly 100 metres long. As might be expected these cables were likely to be run over by vehicles and get cut or snagged on sharp objects from time to time. We were always short of cables and as REME had to make the best of those available many ended up measuring only 80 metres or less. With radio control this drawback could be overcome. So could the problem created by certain shop entrances. I can state categorically there is no way anyone can steer a cable-driven Wheelbarrow through a revolving door.

Another handicap from which Wheelbarrow suffered when I arrived in the Province was its unsatisfactory TV system. Though excellent in bright sunlight, it had very poor capability in low light. The small cameras used at that time were not always able to withstand the jolting they received and were seldom man enough to remain intact when bombs exploded close to them. I am delighted to say that things improved considerably when the Japanese cameras which had been used were replaced by British-assembled cameras.

One limitation which was not resolved in my tour as CATO was the inability to spin the top hamper that sits on the chassis of Wheelbarrow. The vehicle itself is very simple. The chassis is propelled by a pair of rubber tracks powered by electric motors. The top hamper has a projecting single arm on which equipment is fixed or hung. This hamper can move up or down via two electrically-powered arms but it cannot be traversed. If you want to point it one way or the other you have to swivel it on its tracks and in a confined space this is not possible. If the bottom had possessed a sideways movement capability of even 30 degrees off centre it would have been a tremendous advantage.

On the credit side (and its benefits did vastly outweigh its problems) Wheelbarrow rarely broke down and, because of its low silhouette, it presented a very effective blast profile to a bomb. When a machine was blown up and could be salvaged it was generally rebuilt within 24 hours. From the beginning of September 1976 to the middle of the same month, the Belfast EOD section carried out 140 tasks of which 98 per cent involved the use of Wheelbarrow, and at no time were there any difficulties concerning the reliability of the machine. Once again much credit must be given to the REME craftsmen supporting the bomb teams.

Although unauthorised modifications are forbidden in the Army, ATOs did alter Wheelbarrow to suit their own purposes from time to time, and no one took them to task. Everyone was aware that in the hands of a good operator Wheelbarrow was an excellent weapon.

On one occasion I watched Sergeant C— deal expertly with two bombs in the Malone Road. They had been placed in a chemist's shop and, showing more folly than judgment, the proprietor had taken both of them outside and thrown them away. They lay about ten yards apart, one on the pavement and the other in the road. Sergeant C— arrived soon after the incident, deployed his vehicles, drove Wheelbarrow out of its transporter, discussed the matter with the RUC man on the spot and neutralised both devices within ten minutes.

Things did not always go so smoothly. By the time Sergeant S— had neutralised the bomb hidden in the tool box of a Land Rover in a garage in Belfast* he had run down the batteries in one Wheelbarrow and cut the cable of a second. A captain ATO in Belfast had an even more aggravating experience.

Terrorists hi-jacked a van (containing, of all things, a load of shopping trolleys for a supermarket), placed a parcel inside and locked the door. Then they told the driver to take his

* This incident is described in full in Chapter 12.

vehicle to a particular address. The driver did as he was told but eventually, because he was unfamiliar with the area, lost his way and had to ask a policeman. His explanation that he had a bomb to deliver prompted a predictable response and shortly afterwards Alpha team arrived on the scene.

The hi-jacked van presented quite a problem. The ATO quickly realised that he could not get into the back of the van unless he blew a hole in it quite high up – the trolleys filled the lower part of the body of the van. The answer seemed to be Eager Beaver, the heavy fork-lift device. Eager Beaver was promptly produced and Wheelbarrow was strapped firmly to its forks. At the end of the boom of Wheelbarrow the ATO hung a very efficient explosive charge. The advance of Eager Beaver with Wheelbarrow in its stout arms went smoothly on this occasion and the captain manoeuvred it neatly into position.

Confidently he fired the charge and looked expectantly for the results. The back of the van had opened up all right but, alas, he had forgotten Newton's Laws of Motion. Because Wheelbarrow had no backward momentum or recoil system it had been blasted against the massive bulk of Eager Beaver and hung, concertina-like, a complete wreck. To add to the disgust of the gallant captain, the parcel turned out to be a hoax.

The widespread success of Wheelbarrow led to attempts to find a successor – going up-market, so to speak. To this end a device called Marauder was sent out to the Province in 1977. Sergeant C— had already given the benefit of his experience and had experimented in handling the new robot against simulated bombs in England while another operator was trained to use it so that troop trials could be conducted against the real thing. I insisted that these trials should be carried out in Belfast and arranged that Marauder would be the principal weapon of one of the teams in the City, though a Wheelbarrow would travel with them in case there were problems. After training, a warrant officer was selected to deploy the machine and, because it was impossible for me to add the trials to my other commitments, another lieutenant-colonel was chosen by the RAOC for the task of evaluation. He was an excellent choice: not only was he a Weapons Staff Officer, but he had served in the Operational Requirements branch of the Ministry of Defence and been Chief Instructor at the Army School of Ammunition, where all of our ATOs receive part of their instruction.

Marauder proved to be an interesting piece of equipment. With three sections of track on each side instead of the conventional single track it could walk upstairs for a start. It could operate in a confined space and could pivot its top hamper either side of the chassis. It had two separately controlled arms, on one of which we could fix disrupters, a shotgun or other devices. The other had a gripper hand which was so effective that it could open a car door with a key by remote control.

The machine was thoroughly tried over some eight weeks or so and undoubtedly did all the things required of it in the original specification. It was a brilliant design and a mechanical masterpiece. However, independently both the lieutenant-colonel tasked with evaluation and I came to the same conclusion. Despite its many qualities it was unsuited to our role in Northern Ireland at that particular time. Broadly speaking, it was too sophisticated. Like all complex machines it was susceptible to damage from blast and required considerable logistic support in terms of spares and manpower. From 4 May to 17 June 1977 the Belfast Section had 220 tasks. Marauder was deployed on 113 of them. The equipment was used on 45 support tasks, of which 26 were dealt with successfully. With one or two exceptions the tasks it attempted could have been carried out by Wheelbarrow. Finally it was very, very expensive – we could have had four Wheelbarrows for the price of one Marauder. Further trials have been carried out in Northern Ireland, but Marauder is still in the wings waiting to be called on. It may well be considered useful for the Metropolitan Police, the Army on the mainland or abroad, but did not prove itself capable of replacing Wheelbarrow in Northern Ireland in 1977.

However, despite the affection in which it was held by all users, Wheelbarrow was not the total answer to our problems. It was merely the means of giving the ATO a close-up view of the bomb without having to approach the device, and of delivering the means of disruption. In addition, of course, *en route* to its target it could open house doors, car boots and doors and blast holes through the sides of vehicles. It could tow suspect vehicles under certain circumstances, too, but basically it was a weapons system.

The weapons and explosive charges used on Wheelbarrow (or manually for that matter) were developed by another government research establishment. It was a side-line for them, a break from working on guns, shells, mines, grenades, demolition explosives and other forms of ammunition, and the variety of things which they produced for us to trial was limitless. Nothing was too much trouble for them if we had an idea or requirement. The efficiency of their products is well illustrated by the number of bombs that were neutralised. Of those we got to in time, during my 14-month tour, we had a better than 90 per cent success rate ... more than 500 devices ranging from small cassette-type incendiaries to really big stuff.

One snag about the plethora of devices produced by these dedicated people was that ATOs could not try out all of them in England before beginning their operational tour. Their first opportunity was often against the real thing. Thus it was that a tall, fair-haired Cockney staff-sergeant and I had a slight difference of opinion on a winter's day on the main Belfast – Dublin railway line near Lurgan. A dustbin, believed to contain a bomb, had been placed by the side of the permanent way in a cutting.

'We've got a bit of a problem getting down the embankment to the lines,' the staff-sergeant informed me, and explained how he intended to make an indirect approach. The quickest and most obvious route would have meant a full clearance operation with a Royal Engineers search team sweeping the approaches for booby traps but the ATO wanted to open the line as soon as possible.

I was slightly puzzled. 'OK, Staff. You want to do a quick job, but once you're on the line what are you going to do then?'

'I'm going to use Flatsword,' he said determinedly.

I was taken aback. 'Isn't that a bit drastic?'

Flatsword was a large piece of steel about one foot wide and two feet long cut to a point in front. Fixed to a stand it was fired with sheet plastic explosive and had been developed to be fired remotely to cut open beer kegs.

'Surely it would be simpler to fire a shotgun at the dustbin and see if it will topple over?'

The staff-sergeant looked disgusted. 'No way, sir! I'm going home next week and I don't intend to do so without firing a

Flatsword. It's about the only thing I haven't used.'

I could see no point in arguing and we took shelter while the staff-sergeant set up his equipment.

The results were spectacular. Flatsword sliced through the bin as if it were tinsel and a pillar of rubbish shot into the air. As the litter began to flutter down onto the track and it became clear that the whole thing had been a hoax the staff-sergeant turned to me with a gleam in his eyes.

'Wasn't that beautiful?' he said.

When I left, having decided that a tactful withdrawal would enable me to avoid any awkward questions about the appalling mess on the line, he was still muttering contentedly.

Another branch of this particular research establishment which went to great pains to give us the protection we needed was the Applied Physics department. From them came the means of indicating the presence of radio-controlled devices. Their system provided us with the only two successes we enjoyed against such bombs in my 14 months, both at Dungannon.

Much simpler but absolutely essential bits of kit were the Allen hook and line and the bomb suit. According to the rules, all operators had to wear the suit if they were required to walk towards a bomb. It weighed 60 lbs when the chest and abdomen armour was fitted, with a pocket for a radio so that every move could be explained and monitored. Actually, there were times when it was impossible to wear a suit – such as when dealing with a bomb in a culvert. But at all other times it was a good bolster for morale, and I reckoned to wear it for three reasons – first, it set a good example to others; second, it would give some protection, from both fire and blast, if a bomb went off as one was walking towards it; and last, if not least, the time it took to put on always delayed the moment when one had to start walking!

The Allen hook and line was the item which had seen the longest service in the Northern Ireland 'golf bag'. It was always better to move bombs at the end of 100 feet of line than at arm's length. More than one ATO has died handling a bomb he thought he understood when he could have done the job safely with this indispensable piece of equipment. For my own part, I used the hook and line in almost every clearance in which I was personally involved and my most unpleasant moment was at Flax Street Mill when I realised I could not make full use of the Allen kit to clear the booby-trapped tanker.

Despite the fact that the ATOs' armoury was well-stocked by 1976, the research and development establishments never ceased to try to anticipate the next moves of the terrorists in the field of new bomb types and bombing techniques. Their difficulty was making sure they were in possession of the latest intelligence about actual bombing – it came as a shock, to operators and boffins alike, to discover that one piece of equipment had been developed to tackle a type of device which had not been used for three years. Fortunately such errors were rare, and teams flown into South Armagh by helicopter were able to take the essentials with them in two or three Bergen rucksacks – explosives, disrupters, Allen lines and cartridges for the shotgun which was carried in addition to their normal arms. Even the electronic detection kit went in by air in bandit country.

Among the most popular occupants of helicopters carrying bomb teams were the sniffer dogs. These friendly creatures seemed to enjoy the flights, possible because they had long sight and found the view from the open door of a speeding Puma highly entertaining. Every single one of them was slightly overweight because the soldiers insisted on sharing their jam and marmalade 'butties' with them. Because of that, it always struck me as slightly 'off' when we ordered these portly animals to jump through windows which you intended to use to enter a building. This was done so that the weight of the dog would set off any pressure-activated bombs that had been planted to catch the unwary. One's consolation was the knowledge that a dog would not leave behind a widow and children. I am happy to report that despite a considerable number of window entries we never lost a dog that way during my service in the Province.

Outside South Armagh the team operated by road in 30-cwt vans protected by 'plastic armour' and modified with up-rated engines and strengthened suspension. The armour was capable of reducing the effect of small arms fire.

The Pig – the 1 ton armoured Humber troop carrier, did not have the speed of the 30-cwt van, but was a doughty vehicle for use in Belfast and Londonderry. It provided adequate protection against bullets, bricks, bottles and manhole covers unless the occupants were foolish enough to lean out to get a closer look at what was going on. As the teams had to conform to the traffic regulations, such qualities were appreciated when halted at traffic lights in unfriendly areas. The Pigs were driven by men of the Royal Corps of Transport and a good job they did. An RCT sergeant lived with the EOD teams in Belfast and supervised the drivers and the condition of the vehicles, as well as driving the Eager Beaver when required.

One aspect of the bomb scene that no equipment or aid seemed to be able to alter was the attitude of the public, which was utterly unpredictable. The number of people who got into serious and sometimes fatal trouble through sheer curiosity runs into scores. It may be as well to outline certain basic rules to be observed if you have reason to believe there are bombs or bombers about. First of all, be aware. Look out for things that seem out of place or incongrous, perhaps a shopping bag left in an unlikely place or even an unidentifiable smell. Next, remember that if what you have discovered is a bomb, as far as you're concerned it is due to go off there and then. It is not brave to carry a bomb out into the street – it is foolhardy. Terrorists watch to see what action is taken on the discovery of their devices. If someone gets away with moving a bomb once, you can be sure there won't be a second chance. If you see something suspicious leave it and tell the police. If it is a mistake then everyone will be relieved. If it is not they will be grateful.

If you know for certain there is a bomb on your premises, then leave access for the EOD team. It is always tempting to evacuate the premises and lock the door behind you. But if there is a bomb in there and you have sounded the alarm it is unlikely that anyone will wish to make closer acquaintance with it. If the door is locked it will be an added obstacle to the operator trying to get in. The EOD team can, of course, but they will cause damage. Wheelbarrow has a loud knock and no finesse when it comes to leaving a calling card.

Whether Wheelbarrow would be used as extensively in the prelude to a major war is debateable. EOD teams, complete with equipment, are maintained by all three services. The Royal Navy and the Royal Air Force tackle devices found near their installations and the Royal Engineers have a team in Kent.

The main EOD capability would, however, be provided by the RAOC. The teams, fully operational in peacetime, are located throughout Great Britain as part of the No 1 Ammunition Inspection and Disposal Unit. A reserve unit also exists, plus a reservoir of – operators who have Northern Ireland experience, reservists like myself, and the police bomb teams. It is significant that in the home defence experience code name Square Leg in 1980, the scenario included teams of saboteurs and subversives who attacked simulated vital installations with explosives. EOD teams had to deal with their handiwork where they penetrated defences.

My own view is that in a real situation the hook and line, the bomb suit and the disrupter would be the major pieces of equipment used. The emphasis is hardly likely to be on obtaining forensic evidence for criminal prosecutions but the quickest clearance possible for the maintenance of essential services. With this in mind it might well be a case of getting the device away from the target and destroying it immediately, regardless of the noise. That sort of attack against the bomb might be difficult to sustain for any period of time, because constantly walking up to devices would eventually tell on operators. But it might give the necessary breathing space. One thing is certain – the ATO is always going to be among the first into action.

CHAPTER EIGHT

One More Tanker and a Trap

Nobody ever loves a big headquarters. That is one of the facts of military life. Thousands of words have been written castigating the staff officers who served at Haig's HQ at Montreuil-sur-Mer in France during World War I, although these unfortunate individuals regularly put in a 14-hour day and, at times of crisis, went on much longer. I have heard veterans say that the atmosphere at the headquarters of the BEF before the balloon went up in 1940 was chill and terrifying to young visiting officers. And there are a fair number of old soldiers around today who will tell scathing stories about the luxurious, constantly expanding, empire-building Allied HQs in North Africa and North-West Europe.

Headquarters Northern Ireland (generally referred to as HQNI), though minuscule by comparison with any of the establishments I have mentioned above, also came in for its share of criticism. It was only natural that young officers and soldiers worked to the limit during a four-month tour, in improvised accommodation in a disused factory in Belfast or an ancient police barracks out in South Armagh, should blame the distant powers-that-be for discomforts, tiresome regulations and administrative shortcomings. Battalions on 18-month tours with their families, living within all the necessary security restrictions, could also use '. . . those bloody people at Lisburn . . .' as an excuse if things went wrong.

What many of the critics forgot was that the staff of HQNI were professionals like themselves. There was a tremendous amount of experience to be found there, and though World War II ribbons were scarce there were plenty of officers who had seen service in other parts – Korea, Malaya, Cyprus, Borneo, Kenya and Aden as well as the streets of Belfast or Lurgan. Furthermore there was an intimate knowledge of the actual personal hardship being suffered by underpaid soldiers and their families at that time, particularly the enormous heating bills faced by youngsters given quarters which were heated by electricity only. An unrelenting battle was being fought to improve the soldier's lot during my tour and I like to think that the cases presented by Colonel Malcolm Cubiss, the Colonel AQ (that is the chief staff officer dealing with administrative, disciplinary and supply matters), were instrumental in convincing the Government to pay more.

Not all the argument in the world, however, will persuade soldiers from the out-stations that HQ wallahs do not live in the lap of luxury, and sometimes visitors to Lisburn were hard put to conceal their feelings. Oddly, the putting green and the amateur dramatic society seemed to get furthest under the skins of the warriors in combat kit with mud on their boots.

The Ops staff, of which I was one, worked long hours as well. A 12-hour, 6-day week was the norm when things were quiet.

I must confess that the gravity of the struggle on the putting green quite startled me when I arrived at Lisburn. The course itself was regarded as a piece of ground to be cosseted and protected and there was always a section roped off to recover from the ravages of the fray. No one ever forgot to shut the knee-high gate leading onto the combat area. In the scorching summer of 1976 the sward resembled a rather tatty brown carpet most of the time but was still accorded due respect. Long before the league table results were published in Visor, the service newspaper for the troops, the rumour would go round that one team or another had put one over their rivals. As the struggle intensified one would occasionally see a group of four thoughtful figures carrying putters (yes, even during working hours) heading for the northern end of the camp to do battle with their opponents. There was no distinction where rank was concerned – such was the shortage of staff that those who were present made up the teams whether they were good at the game or not. Great was the glee when unfancied teams pulled off a surprise. This was not unusual. It was regarded as bad form to practise.

For some reason I could never fathom, the Special Investigation Branch were the putting experts, though it probably had something to do with the subtlety (or plain cunning) of their leader Major Stan Upton, a man of aldermanic

The author (*right*) inspecting the firing platform after the mortar attack on Crossmaglen

The protective netting at Crossmaglen which stopped the mortar shells exploding

Lieut-Col Patrick with the PIRA mortar bomb Mark 9, and a 'Pig' in the background

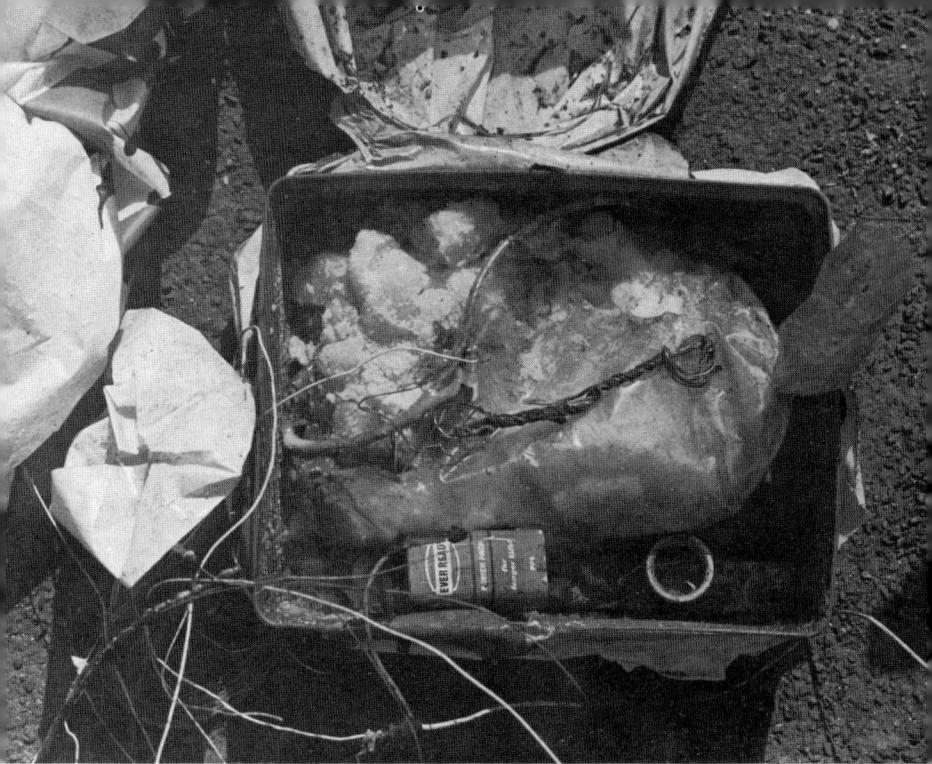

What's inside: this bomb contained 15lbs of CO-OP and a battery power pack. It was blown open with a disruptive device

The remains of a car in which an RUC constable died. He was leaving for work; his wife was waving goodbye

The author (*nearest camera*) and members of the Royal Engineers search team pulling a bomb into the open

A 'come-on'. The car had been deliberately blown up to attract the
security forces, and the churn on its side was booby-trapped

Inside the booby-trapped churn, which contained 94lbs of explosive

Lieut-Col Patrick on top of the tanker which forced the decision to clear the booby-trapped car. The snatch block was used to open the lids remotely

No joke really. A few seconds after the EOD team ran for it, the bomb in this pub exploded

appearance and inscrutable countenance of whom it was said that he had never been seen in uniform. It was even alleged that Stan did not possess a uniform, though I know that at least he owned full mess kit for I had seen him in it – complete with 13, or was it 14, miniatures recalling campaigns going back to the time when he was a young Army commando at the Rhine crossing in 1945.

The drama group seemed to be the prerogative of the Educators and the Intelligence Corps. It created a strong interest, also, for the wives of officers living on 'The Patch', and was strongly supported. They normally performed plays, such as a thriller in 1976. But the following year, inspired by Major Shirley Neild, the senior Women's Royal Army Corps representative at HQ, the Harp Players went out on a limb and decided to do 'The Pirates of Penzance'. For a time it was extremely risky to enter the mess bar for fear of being buttonholed by Shirley whose warm personality and winning ways had succeeded in 'enrolling' the most unlikely cast, some of whom had sung only ribald soldiers' songs previously.

There were plenty of other relaxations too, with great scope for golfers, even though the IRA occasionally attacked clubs used by members of the Security Forces and blew up the professional's shop at Malone shortly after the Ulster Defence Regiment had a meeting there.

Cricket was also popular, though many a match was interrupted by the arrival of a helicopter at one end of the pitch, whereupon all the players moved out of the way until the coast was clear, as if it were the most natural thing in the world. I have even played in a match where the arrival of a padré by chopper was said to have been a pre-arranged plot to break the batsman's concentration. The Officers' Mess team was enthusiastically led by Lieutenant-Colonel John Williams, Army Air Corps.

The Mess itself was comfortable. Once the home of a linen mogul it offered a certain sombre Victorian comfort and I occupied a sort of suite created from a divided room. As the numbers had grown, a wing had had to be built onto the old house and then huts, linked by a corridor to the wing. At the bottom of the hill, near the hallowed putting green and the MT sheds, stood yet another annexe.

The original owner of the estate seemed to have been a man

of taste for the grounds were dotted with noble trees and fine rhododendrons, which softened the appearance of the red brick main office blocks in which the staff worked. Houses for senior officers, including the Commander Land Forces, formed a neat estate.

The perimeter wire of the camp also enclosed married quarters for other ranks, with a NAAFI, laundrette and school, plus Thiepval Barracks, with its soldiers' blocks and the headquarters of 39 Brigade.

All in all the place had a civilised air, and the armed guards and dog patrols merged with the scene after a while. Housewives with babies pushed their prams into Lisburn giving only a cursory glance to the noticeboard on which the 'State of Alert' of the Province was recorded . . . from black, to amber to red.

The housing estates around the camp looked surprisingly well-to-do and it was not unusual to hear the troops pointing out that there were far more new cars parked in the drives and streets than in their hometowns in Great Britain.

It was a strange atmosphere in which to be fighting a war and, on a fine day, the idea seemed utterly unreal. The reality was there, however, and on the doorstep. A sullen 'crump' one lunchtime in the winter turned out to be a car bomb exploding in Causeway Park Estate in Lisburn which was occupied solely by officers and their families. About 15 minutes walk away, it was isolated by road works and fields. Parked at a T-junction the bomb caused only superficial damage to the surrounding houses and no one was hurt, but only by a chance did it fail to catch a group of children who normally came along about that time: they were a couple of minutes late that day. A lighter note was struck by Felicity Stockton, wife of the editor of Visor. She had been in the kitchen when the bomb exploded and she confessed that her first reaction had been, 'What has that damn dog done now!' For once, however, the boisterous golden retriever was not guilty.

Another bang signified the blowing up of a hotel only a few hundred yards distant. The place was evacuated before the explosion but considerable excitement was generated when it was discovered that a Canadian visitor was missing. He was discovered sleeping blissfully in the wreckage . . . protected from grim reality by his total deafness.

Perhaps the most telling reminder of the true state of affairs was to be found in the watchkeepers' room of the Army Information Service, which, at that time, ran a 24-hour service for the media.

It was normal for me to pop in from time to time to keep in touch with things, perhaps to brief the Chief Information Officer or to check something with one of the watchkeepers. The Press Desk monitored and recorded all television and radio news bulletins, and it was possible to pick up a lot from them. Someone had posted a notice on the TV set: 'Will the last watchkeeper to leave Ulster please switch off the Telly'! One side of the room was completely covered by an illuminated street map of Belfast with a ready reference system so that the location of any incident could be seen at once. All military maps of Belfast were in 2 colours – Green and Orange. Green for Catholic areas and Orange for Protestant. It became second nature to record automatically in one's mind the colour of the area in which an incident had occurred.

At busy times, stars, blobs and question marks dotted the map, indicating the whereabouts of shootings, bombings, and other incidents, black if no one was killed, red if there had been a fatality. A similar map on an adjoining wall gave a similar state-of-turbulence picture of Londonderry.

To the right of that was the running total, a board giving the statistics of the campaign commencing in 1969 when the total deaths, civilian, military and police totalled 13. The record then climbed dismally – 25 in 1970, 181 in 1971, a staggering 482 in 1972. The following year saw a significant fall to 247, dropping still further to 217 in 1974; but the 247 figure was reached again in 1975 and there were nearly 300 in 1976, which turned out to be, unhappily, the second largest casuality list death roll of the ten years' troubles when the count was made in 1979.

All sorts of data could be gleaned at a glance from the stats. board including, at the bottom right hand corner under the tag 'Latest Fatalities', the names of the last people to die. There were four columns – Army, UDR, RUC and Civilian. Sometimes, if there had been a dual or triple killing, there would be more than one name in the appropriate column. You could not help looking at these whenever you visited the room. Someday the name alongside the Army tag might be yours. It would remain there for a time and then be replaced by someone else's. Bernard Calladene's name had been up there at one time, and Barry Gritten's and Gus Garside's.

That was how the war came home, at the same time both personal and impersonal.

Many times, of course, the casualties were unknown to us, but often, in the case of a military death, there was someone in the Mess who had either served with the victim at some time, or knew the unit, or had visited it or dealt with it at some time.

Occasionally grim details that were not made public filtered through and might be mentioned in the bar of the mess ... 'You know that UDR man they got as he was getting into his van the other day. Well, they didn't kill him first time. They had a youngster with a shotgun with him and they sent him out to 'blood' him. The little bastard put the gun to the UDR man's head and blew it to pieces.'

'I hear that there were only slight injuries when the bomb went off...' 'It depends what you mean by slight. It looks as though the corporal is going to lose the best part of an arm.'

Yet the mess itself, and the bar in particular, was primarily a cheerful place, often perversely so at difficult periods. As it ran a snack bar service it was frequently packed at lunchtime, and the crowd might include visiting police chiefs, civil servants (including some pretty secretaries from Stormont), an odd newspaperman in the care of the PROs, and a clutch of officers from various units, perhaps in Lisburn for a conference. All regiments and corps were to be seen there at one time or another.

On the other side of the counter worked the best barmen I have come across, all local civilians from what were regarded as 'dodgy' areas who gave as good as they got in the banter that went on.

'I'm tellin' ye, sor, there are only two kinds of officers in this mess.'

'And what's that?'

'Thim with dogs and thim without dogs ... and I'll tell ye whose side I'm on if ye tell me whether or not you've got a dog.'

The dogs were a constant source of controversey and were frequently the object of barring regulations. All belonged to an illiterate breed and took little notice of whatever orders were published affecting them. To make matters worse, there was a perpetual rivalry to establish just which of them was senior dog in the mess. The otherwise impeccably behaved labradors, one black and owned by a Royal Welch Fusilier and one golden and owned by a 14th/20th Hussar, were streets ahead of the other canine inhabitants but refused to give way to each other. From time to time occupants leaped for the safety of the bar or the tops of tables as the competition boiled over and two growling combatants were pulled apart by their cursing masters, both of whom were excellent soldiers. The contest was eventually solved in a sad way as the golden labrador was run over not far from the mess, and died of his injuries. His master,I might add, was decorated later for his work against Loyalist murder gangs.

Dogs included, the mess was run on definite peacetime lines and you had to turn up promptly for meals otherwise you might well go hungry. The watchkeepers in the Ops room at HQ 39 Brigade, who worked a 24-hour shift system, seemed to miss out most when it came to eating and it appeared to me that when they were on nights they went either short of food or short of sleep.

Contrasting the niceties of life in the mess with what went on elsewhere, it struck me now and then that this was a very strange way to be at war.

Even stranger was it to come home to London. A 20-minute drive from Lisburn led past signposts pointing to places where there had been murder and mayhem; sleeping policemen (tarmac bumps in the road) forced cars to slow down as they approached the checkpoints manned by men of the Royal Air Force Regiment; and you walked through a metal detector and presented your luggage to teams of searchers before you were allowed into the departure lounge at Aldergrove. The luggage went through its own security route to the plane and a great deal of checking and counting went on before take-off.

In London, apart from the Special Branch men at the terminal, there was nothing to indicate there was any problem at all only an hour and five minutes' flight away ... no road blocks, no oil drums along the main thoroughfares ... no netting outside the police stations and no slow-moving Land Rovers with a soldier's head sticking out of the top and a rifle at the ready. All the cafés and pubs one passed were accessible if required. It was safe to talk to people – and people didn't mind a casual conversation.

The longer one stayed in Northern Ireland the more one noticed the difference, especially if something happened in the Province during one's temporary absence in England. 'Something' did happen on 24 March while I was at a Royal Army Ordnance Corps Ammunition Depot, giving a lecture on the situation in Ulster. The Provos repeated their attack on the Donegall Pass RUC station, this time using a diesel tanker. Before it could be neutralised the device detonated, setting the fuel ablaze and doing considerable damage to the station and surrounding buildings.

Four days later I was in the office hut at Lisburn when a 39 Brigade watchkeeper rang to say that a petrol tanker had been hi-jacked and left outside the Kingsway Memorial Hall, at Dunmurry, on the road to Belfast. I shouted for Coupe and clamped our Kojak light on top of the car as he started up (it was one of my toys and I don't deny I enjoyed playing with it. It made people get out of the way).

With sirens blaring we roared out of camp and down the road, and I used my radio to contact the Belfast EOD detachment and HQ 39 Brigade.

'Hello Zero and one five. This is one five Zulu. Am *en route* for incident given by Zero. Over.'

'Zero, Roger, out.'

'One five, Roger, out.'

It was just after 2 pm. Alpha team had been sent out from Belfast and it became clear that once again we were going to end up at opposite ends of the tanker, as in the previous incident at Donegall Pass. This time, however, there was no fog and though we had to take a circuitous route, suggested by an RUC constable, we made our rendezvous with the EOD team only ten minutes after leaving Lisburn. A detachment of the 2nd Regiment, Royal Military Police, had set up an incident control point, laid out the cordon and cleared the houses by the time we arrived and Staff Sergeant Bradder, who was in charge, had the tanker driver on hand waiting to be questioned.

The driver was a level-headed man with a straightforward story. He had just made a delivery and his tanker, a vehicle owned by Petrofina, had been hi-jacked in the Falls Road about

1.30 pm. When full it held 7,000 gallons in seven equal compartments, but No 1, and Nos 4 to 7 were empty. No 2 tank contained 800 gallons and No 3 about 200.

'The baarmb is in No 5,' he said emphatically.

'How do you know?'

'I put it there myself.'

'Oh, yes. . . .'

'They made me.'

'How did you put it in, then?'

'I lowered it on a string, didn't I. It was in a paint tin, a white paint tin . . . about half a gallon. There was some newspaper on the top of it.'

I was glad the driver had been so observant. 'How did they secure the bomb to the inside of the tank after you put it in?'

'They didn't. Once it touched the bottom I let go of the string.'

He was cool. 'And then?'

'I closed the lid on the top of the tank.'

'Did you lock it?'

'That I did, sir. Sorry if that was wrong.'

I didn't complain. The driver had been placed in an impossible position by the terrorists who, after making him lower the bomb into the tanker, had ordered him to drive to the city centre. He had got as far as Dunmurry when he decided he'd had enough and got out.

Just as I had finished questioning the driver, up drove S—, the officer commanding 321 EOD Unit, with Lieut-Colonel Mike Newcombe who was on one of his frequent liaison trips from Didcot. They had heard about the incident on the radio while driving back from a visit to Lurgan. In fact, this might well have been S—'s bomb, as after the Donegall Pass affair we'd decided he could have the privilege of sorting out the next one.

Now that I had begun the operation, however, I said I would go on with it. I borrowed the bomb suit belonging to Sergeant M— leading Alpha team and then we all had a cup of tea while we sorted out the equipment that we would need.

The drill appeared to be straightforward. I would have to walk to the tanker, climb the ladder, open the lid of the tanker and fish up the bomb. The device then had to be lowered to the ground and neutralised. Sounded simple enough.

The bomb team's kit contained a series of short rods that screwed together like a chimney sweep's brushes. We made up an eight-foot length so that I could reach the bottom of the tank.

The day was clear and dry, but once I began my trek to the tanker I wouldn't have noticed if it had snowed. In my left hand I clutched the rod, in my right the disruption device. Despite the gaggle of experts who were watching me there was no point in looking back. I simply kept on heading for the tanker, watching the letters 'FINA' growing larger and larger.

I left the disruption device on the pavement and then began my slow ascent up the ladder, once again a most difficult climb in a bomb suit.

Once on top I went straight to No 5 tank and unlocked the lid with the driver's key. The rod I laid beside me. The bomb was just as the driver had described it. The light was good and I had no need of a torch to see it lying on its side at the bottom of the tank with a splash of white paint beside it. The handle, to which the lowering string was attached, was clearly visible and I had little difficulty in reaching it with the hook on the end of the rod. Carefully, hand over hand, I brought it up and than walked to the side of the tanker to lower it. Problem! The rod which had been long enough to reach the bottom of the tank could not reach the pavement . . . naturally enough. Now what the hell was I to do with the thing? I bent forward as far as the heavy bomb suit would let me without plunging over the side and froze in a position which one of my old sergeant-majors used to call a 'constipated duck'.

To stay in this ludicrous position for ever was clearly impossible. Had I felt flexible enough, I could of course have pulled back the rod and carried the bomb down the ladder in one hand — but it didn't strike me as a great idea. I then reasoned that if the paint tin has been driven from the Falls to Dunmurry on its side, it must have rolled about considerably. Conclusion: there was no tilt switch and whatever else was inside was fairly robust. If I dropped it the remaining foot or so the chances of it being activated were amost nil. It was just possible that the time set for detonation had been reached but had failed to work because of a poor contact. A jar might theoretically re-establish the contact, but the possibility seemed remote. So I dropped the thing, hook, rod and all. It clattered to the ground and that was all. Just a clatter.

It was an action I would not expect to be taken by any but the most experienced operators and no training manual should recommend it. I can say only that deductions were right in the given circumstances.

Closing the lid of No 5 I retraced my clumsy steps down to the street, unhooked the device from the rod and lined up the disruption device before retiring to the safety of the Pigs and the incident control point 100 yards away.

After connecting the leads I told the radio operator to warn 39 Brigade HQ that there was going to be a small – I hoped – explosion, called out 'Firing' and pressed the button.

The bang was almost delicate and when I went back to the device I found it was neatly separated. I called forward two of Alpha team who examined the remains and took samples for forensic purposes. The bomb had been made up of 5lbs of CO-OP, a timing and power unit with a Parkway one-hour timer, a short delay detonator (American made), and it had been armed with a clothes peg switch. The wooden dowel which had held the jaws of the peg apart had been withdrawn to start the timing circuit before the tanker driver was told to lower it into the vehicle and easily might have gone off while he was driving. If anyone could consider themselves lucky, he could.

From the time I arrived on the scene to the point at which I was able to declare the area clear two and a quarter hours had elapsed, though the time I had spent in dealing with the device amounted to only 20 minutes. As I washed my hands in the washroom of an Esso station not far from where the tanker was parked – the staff seemed delighted to afford me the facilities – I mused over the tactics used by the IRA on this occasion. It seemed certain that they would not take this lying down and we could expect another hi-jacking in the near future. I felt even more certain that this was only a chapter in the series when the next day *The Belfast Telegraph* carried a picture of me at work on the tanker, together with a description of the operation. This was the second time I had 'appeared in print', as the Donegall Pass affair had also been widely reported.

The question of media coverage of bomb incidents is a vexed one in Northern Ireland. By and large the Press and the television crews know the ropes and will not knowingly put out pictures which enable members of EOD teams to be identified.

Everything possible is done to prevent individuals being set up as targets. In my case, however, as I had to brief correspondents and appear 'on the box' for our own purposes, there was no real reason for me to object to being filmed – except that the more a person was talked about, especially where successes against the IRA were concerned, the more reason they would have for trying to eliminate him. To get rid of the Province's chief ATO would constitute a great feather in the cap of whoever pulled off the coup.

This was what must have been going through the mind of Colonel Cubiss when he saw me at headquarters the following day.

'You'd better watch your step in future,' he said. 'Next time they do it you may not be so fortunate.'

The 'next time' the Colonel AQ had been so concerned about arrived on 18 April. As usual I received the news by telephone in my office at Lisburn – at precisely three in the afternoon. From the beginning the whole thing had a different feel about it. Something deep down was telling me to beware. I didn't have the same approach as to the previous tanker incidents at all. There was no rushing out to clamp the Kojak light on the roof and we did not use the sirens until we were almost at the scene. The Crumlin Road, leading from the city centre past the jail and with the Ardoyne and other choice areas at hand, is not the place to draw attention to yourself. For all I knew my car might have been compromised – identified and listed by the IRA – and a sniper could have been waiting for it.

The tanker was a Leyland-built Shell vehicle carrying diesel oil – much of its cargo of 2,750 gallons had been discharged but though the first two tanks were empty it still contained a thousand gallons of fuel in the back tank. It had been hi-jacked in the Crumlin Road area about 11.30 am and had been parked outside the Security Forces base at Flax Street Mill just after 2.30 pm.

A major in the 1st Battalion, The Devonshire and Dorset Regiment, was at the ICP in command of the incident and he gave me a full brief when I arrived.

The 'D and Ds' are a regiment with considerable experience of Northern Ireland and a high reputation. When the tanker had appeared outside the mill the sentry had wasted no time aiming his rifle at the driver and giving the explicit instruction:

'. . . . off. Get that . . . . ing vehicle away from here and down the . . . . ing road.'

Such unequivocal language was not wasted on the driver and he had moved the tanker some distance before abandoning it.

'The poor sod's just back from work after recovering from a heart attack,' said the D and D major. 'He's in a right state. The RMO is looking after him . . . they're in the Saracen over there.'

I asked the major to let me speak to the regimental medical officer before I spoke to the driver.

'He's in a bit of a state, poor chap,' said the RMO. 'He's only just gone back to his job after a long lay-off. They held him for three hours – in a room somewhere – while they messed about with the tanker. Then they took him back to the cab and told him to drive slowly and carefully over here and leave the tanker. Said he wasn't to bounce over any rough bits of road – I ask you! He's about all in.'

'Can I talk to him?'

'Sure. I've given him something to calm him down. But keep it short.'

Sitting in the back of the Saracen in his dungarees, the driver didn't look too good. Someone had given him a mug of tea and he was sipping it and staring over the rim. A couple of his mates had turned up. He shuffled up a bit as I sat down beside him and did his best – I think he realised that we both had problems but that mine was likely to be solved a great deal faster than his if things went wrong.

After describing his long wait under an armed guard he said, 'I think there must be two bombs. I think they put one in the front tank and. . . .' He took another sip of tea. . . . 'They left a box of tricks in the cab with me. It was there as I drove along.'

He cradled his mug in his hands and sniffed. 'I drove careful all right, sir. I'm not a well man.'

'Well, you've been very helpful. I hope you'll be feeling better soon.'

He looked at me with hollow eyes. 'Thank you . . . and good luck.'

'Thanks!'

I walked away from the Saracen with the realisation

dawning that I was very frightened indeed. This was one bomb I did not want to deal with.

Captain T—, who led the Belfast section at the time, was commanding the EOD team on the spot so I walked over to him to discuss the situation. The usual people were beginning to appear from the public utilities. We arranged for sand to be delivered and a 'bund' to be built across the Crumlin Road to prevent oil flowing into the sewers should it escape. A foam vehicle from the fire brigade was placed on standby close at hand. Of the three tankers I dealt with in Belfast this was the only time I actually put down sand. That showed my feelings about this device.

While we were talking the Deputy Commander of 39 Brigade, Colonel Eccles, arrived and the OC of 321 EOD Unit turned up with the lieutenant-colonel who was on a familiaris-ation visit from Berlin (to take over from me later in the year). My admin officer, an avuncular old soldier who for some reason I never understood went under the nickname of 'Sniffer', was also in the party.

The options open to me were the same as those on the pre-vious tanker clearances. I had to get onto the vehicle, open the filler cap and take out whatever was inside. The other device I was fairly certain must be of a secondary nature – the terrorists were out to create a powerful explosion using the medium of the fumes in the empty tank and not merely to wreck the cab.

About half past four while the sand was being put down by an armoured RE bulldozer, I decided to get on with it.

'Right, Geoff, got your bomb suit?'

He nodded.

'Nice and clean is it . . . then I'll borrow it.'

I tried to sound cheerful but I don't think it came off. Captain T—, who was later awarded a Mention in Despatches for his work in Belfast, could sense that I was on edge. The strain was beginning to have its effect.

After the body armour had been fitted the OC 321 EOD got some strips of black jungle tape and began sticking them on the front of the bomb suit so that if I needed them in a hurry they were in the most convenient place. He tried to ease the tension by saying, 'Well, Colonel, they'll help to hold you to-gether if the bomb goes off!'

That remark did not go down at all well with Captain T—

whose face set as he asked me, 'Do you let people talk to you like that...?'

Quite honestly, I was more concerned with the black tape than the remark, but I replied, 'Not normally, but in the circumstances I've got other things to worry about.'

We began sorting out the hook and line kit and after explaining to Colonel Eccles what general precautions had been taken for the safety of the area, I set off.

It was then 5.30 pm and T— walked part of the way with me while I told him my intentions.

'If I can fix the hook to the bomb and get the line above it somehow we'll try to pull it out remotely.'

'Right you are, Colonel, we'll wait out until you give the signal.'

Almost as an afterthought he added the familiar phrase, 'Watch your step, sir.'

I walked towards the tanker wishing I was somewhere else. I went up the steps at the side of the tanker full of dread, because this time I had no idea what the IRA had done.

The bomb was in a plastic putty bucket. The handle had been slipped over a piece of wood and this had been jammed across the mouth of the front tank. I taped the hook and line onto the bucket handle and looped the line over the top of the filler cap. In this way I hoped that I might get sufficient leverage to pull out the device. The only thing that kept going through my mind was, 'Don't touch it.'

We tried to move it— I walked back to Captain T— and we both pulled. Nothing happened. He did not say a word.

'It's stuck under the rim of the tank. I don't like this one little bit. I think I'll have to take it out by hand.'

I was now quite certain that this bomb was meant for me. *They*, of course, knew exactly how the devices had been rigged up in the previous tankers. And *they* had a good idea that as I had done the previous clearances they stood a good chance of getting me onto this tanker. Furthermore *they* could guess my likely movements. They had read the newspapers and, for all I knew, could have watched from a distance with binoculars.

No one was better placed than our opponents to design a trap for me and I felt sure this is what they had done. From what the driver had told me about being ordered to avoid any bumps it seemed certain that the bomb was now live and had been, probably, from the moment he was told to get back in his cab. If the piece of wood itself was not some sort of trigger mechanism – and that was unlikely because of the way it has been jammed into position, the most likely source of danger lay in the bucket itself . . . perhaps a simple tilt switch.

Everything in my body was telling me not to go back to that tanker but I had no choice. I had to go. Dumb piece of metal it might be, but it looked evil and when I put my hands on the cold metal of the rungs leading to the top it was as if a shock went through me. Belfast is full of traffic noise at that time of day but I heard nothing.

The hook was still taped to the bucket handle as I knelt over the opening below the filler cap and studied the bucket. The phrase burned in my brain. *'Whatever you do don't tilt it.'* Next I leaned over the opening and felt the underside of the wood. It was rough against my palm but there was no sign of a wire. Using one hand to steady myself I twisted the batten slowly to one side to come away and I took the weight of the bucket as it did so. It was not as heavy as I had expected.

'Whatever you do don't tilt it. . . .'

I gauged the centre of the aperture, which was about a foot in diameter, brought the end of the batten up slightly so it would come through the gap and lifted. The bucket emerged balefully into the light of day, almost resentful that it had not bumped the edge of the hole. Slowly and deliberately I set it down at the top of the steps, leaving the piece of wood through the handle and resting on top of the bucket.

After a thorough inspection to make sure there was no way I could catch the line accidentally, I began my descent.

At about the third ponderous step my eyes came level with the bucket and I found myself staring at it with loathing and fear.

'Whatever you do don't tilt it. . . .'

On my first foray I had walked back to the incident control point 100 yards away. On this occasion I wanted to get away from that bomb as quickly as possible. For that reason I moved about 20 yards and turned the corner of Flax Street into the Crumlin Road. Just being out of sight of the thing made me feel better. I gave myself a breather then picked up the cord that led to the hook taped to the bomb and pulled.

Fishermen will know the feeling of a line that goes limp as a big one gets away. For the tiniest fraction of a second I had a similar sensation. Then there was an almighty crash and blast of hot air. Rubbish whirled around me and torrents of broken glass poured from the windows of the Crumlin Road, jangling and smashing. I flinched as something bounced off my unregulation lightweight helmet, and stood crouching stupidly with the bit of slack line in my hand.

I'd tilted it all right. The bomb had been rigged as I expected and, falling from the tanker as I pulled, had exploded when level with the cab window.

The faces of the party at the ICP were, if anything, paler than mine when I made my way back to them. They had been talking to the CO of the D and Ds when I left the lorry and had not seen me as I went up the hill away from them. When the thing went bang they looked round and couldn't see me. They were sure I had been killed.

Sniffer just couldn't take his eyes off me as I removed my helmet and wiped my brow. He was the first to speak.

'Colonel,' he said in his honest North Country tones. 'You're a very brave man.'

And a very frightened one ... but old Sniffer's words made me think, for the very first time, that a man does need something extra to sort out the one which is meant specially for him.

Somebody pushed a mug of tea in my hand. I had a quick gulp and then put on my helmet again and went back to the tanker to deal with the second device in the cab which seemed to have suffered remarkably little. The blast had knocked over the cardboard box containing the device the terrorists had left in the cab and it looked as if it was full of rubbish. I was in no mood to jump to conclusions, however. I placed a disruption charge against it and blew it according to the book. As I suspected, it turned out to be a hoax. At 6.10 pm I declared the area clear.

Evidence collected on the spot made me more than ever sure that the bomb had been intended for me. This was no amateurish affair with a home-made blast mixture but had contained 2lbs of powerful industrial explosive – I had suspected that from the lack of smell when I lifted it out of the tanker. The face of the timer, which we recovered, showed it to be a Coral travelling alarm clock, and the power supply was two PP9 batteries. The attack bore all the hallmarks of a specialist operation and I believe firmly that the Belfast IRA had brought in a booby trap expert from Strabane, south of Londonderry, to do the job. Hence the three hours the terrorists had taken to set up a device of such a sensitive nature. I felt rather sore that the Intelligence people had not been able to give me any warning. They should have had some idea.

Another thing that annoyed me about the incident was the strange attitude of the police at Tennant Street RUC station after the explosion. I arranged for the tanker to be driven there to be examined for fingerprints and any other evidence (other than that obtained for DIFS). But they were not interested. Certainly they may have had a lot on their plate at the time, but I was disgusted at their attitude which I viewed in the light of having risked my neck to save their sewers from a flood of dangerous fuel. Thank God they did not reflect the dedication of the rest of that remarkable force.

When I returned to Lisburn about 7 pm I went straight to the bar. It was customary for my particular chums to take the mickey when I entered, good natured banter which I accepted and returned. This was the first time it didn't happen. Pat Lewis, the Royal Air Force wing commander who acted as air staff officer to the GOC, simply said, 'Pint Derrick?' and I found a tankard in my hand in a twinkling. He and John Peecock, the gunner lieutenant-colonel who served as military assistant to the general, were standing with John Williams, the Army Air Corps lieutenant-colonel who flew me regularly to incidents. All were neatly turned out in civilian clothes. A stranger brought in from the darkness for a drink that night could not have guessed from the conversation what our professions were. I may have laughed a lot more than usual and got a lot of strange looks. But no one spoke about the Crumlin Road, though they'd heard all about it. The sensibility of British soldiers, commissioned and otherwise, is a remarkable thing. Intuitively they knew I was undergoing a severe reaction to the events of the day – but understood this was a private soul-searching, best left alone. The homely question, 'Pint Derrick?' had brought me back into the real world, and I intended to stay there.

The following morning John Peecock rang my office and asked if I were going to be in for the next half-hour. The GOC, Lieutenant-General Sir David House, wanted to see me.

Our hut was just behind the main building and we did a quick tidy-up before he arrived.

'How are you, Derrick?'

He was looking at me in his usual calm way, the sort of cool gaze that leaves no room for equivocation or woolly replies.

'I'm all right, General, thank you.'

'How do you feel after yesterday?'

'I still feel a little shaky but I'm perfectly OK.'

The gaze pinned me down still further.

'You've done all that we could ask of you, and there would be no stigma attached if you decided to go home now.'

I was a bit startled by this.

'No, General. Thank you, but really I am all right.'

He paused for a moment, then leaned towards me – we had a chair for visitors.

'Well then ... the next time you are called out to anything like that, you are to let me know. I inisist on going with you.'

If I didn't swallow hard I should have done.

'Right, General. I'll let John know if there is another one.'

With that the GOC smiled happily and left. I took some slight consolation that if I had to have a hand in getting the general blown up at least his MA would share the blame.

I never felt the same towards the job after the Crumlin Road tanker incident. The jaunty approach I had developed was gone, and it took some time before anything like it reappeared. I began to realise just how much I wanted to stay alive and see out the rest of my tour.

CHAPTER NINE

Against the Civil Servants

One result of the Flax Street incident was a meeting of a committee of the Northern Ireland Office to study ways of preventing tankers being used as proxy bombs – a 'proxy' being delivered by some unfortunate individual under threat of duress and not by the terrorists themselves. As I was the person ultimately responsible for the clearance of improvised explosive devices – IEDs – I had to attend. Another of my responsibilities, of which I was acutely aware, was to decide, in the case of a device that did not threaten lives directly, whether or not to put an ATO at risk in order to neutralise it. Because of the need to obtain evidence I had given directions to teams that the destruction of devices was to be carried out as a last resort. Wherever and whenever possible they were to be recovered intact.

In theory, of course, it is possible to avoid all risk to operators by decreeing that they should tackle either every bomb remotely or isolate it until it has gone off. A tremendous amount of damage would be done but no one would be placed in peril deliberately. Least of all, of course, the people who laid the bomb because only minimal evidence would be obtained for possible use against them. This course has never been adopted or, as far as I know, considered seriously.

It is now accepted, almost as if it were in the nature of things, that 'come the bomb, the ATO' and risks of varying degrees will be taken. This reliance on an automatic response to the threat posed by bombers has its drawbacks. Traders, shopkeepers, and businessmen sometimes show a remarkable degree of apathy, complacency or indifference to EOD operators, particularly if their own premises are not menaced. Especially where they are involved personally, they cannot be unaware that they will benefit from the government's compensation scheme should damage be done. (Unfortunately, stories of small traders setting fire to their shops, or even arranging for them to be bombed so that they could claim, were common enough and at least some of them contained an element of truth.)

From time to time some civil servants showed ignorance of our role, a marked lack of sympathy and even downright insensitivity.

As among the members of any profession or section of society, I have found civil servants to be a mixture. Many of them, particularly in the higher echelons, are men and women with good brains and a capacity for very hard work; others time-servers, dedicated but uninspired; a few casual and unhelpful. They are only human, I suppose. There is a small stratum of high flyers, however, perhaps not too long out of Oxbridge or other seats of learning, over-promoted and without experience of ordinary life, who display the worst kind of arrogance.

Around the table at the tanker security meeting was at least one of those. The discussion covered a fairly wide range. One urgent move was to stop drivers carrying the keys to filler caps – at least that would prevent bombers using a gas-filled hull to increase the power of the explosion. We went on to discuss further measures, during which it was made clear that they must not be of such a nature as to put the tanker drivers at risk, and this was quite understandable. Perhaps less so was the insistence of the oil companies that not only must further precautions not interfere with their filling and delivery operations but that any cost arising from them must be borne by the government. It was down to the taxpayer once more.

Perhaps conscious of his duty to safeguard the nation's moneybags, perhaps trying to put the ball back in the oil men's court, one of the civil servants present remarked, 'As far as cost is concerned it is cheaper to have an ATO killed than it is to modify the tanker fleet.' An embarrassed silence followed.

It was a remark too stupid to be callous, but I took immediate and violent exception.

'If that is your view,' I said, 'there is something you should know. If hi-jackers park a tanker near Stormont, or your homes, or your fuel and water supplies, or any other Northern Ireland Office-controlled amenity, I will not hesitate, in the

107

future, to allow it to blow up. It will make life a lot simpler for my men, I can assure you.'

Once again there was silence for a moment before a young but senior civil servant asked me if I meant what I had said.

'You can take it as a policy statement, as far as I am concerned,' I told him.

I was so angry that I found it difficult to sit through the rest of the meeting and, apart from that episode, remember little today about what decisions were taken. Maybe we adjourned – usually we did. Certainly the gentlemen in question were lucky that during the rest of my tour no incident occurred of the nature discussed which would have given me the opportunity to exercise my prerogative either to send in an ATO or wait indefinitely to see if a bomb was live or not.

Yet this was not the only time I was obliged to 'threaten' people on my own side. After one of the Belfast EOD teams had cleared a device in a hotel between Lisburn and Belfast, two police vehicles turned up and detectives prevented them from leaving. They insisted on searching the EOD vehicles.

The police had been called by firemen whom the team had ordered to leave the building as a second device was suspected, and the implication was that the team might have used the opportunity to steal something. Nothing was found, of course, and eventually the team was 'allowed' to return to Belfast – where their mood may be imagined.

The next morning I drove over to the Royal Ulster Constabulary HQ at Knock and told the head of the CID, 'The next time the police do anything like that I will keep my soldiers off the streets of Belfast and you can deal with your own bloody bombs.'

We were good friends and had worked together a lot. He knew I was capable of carrying out my threat and I had no cause to make a similar complaint to him.

Bureaucratic niggling from public agencies and general indifference or obstruction from the man in the street were far from uncommon – particularly in Belfast and the big built-up areas. The population had become inured to bombing and drivers rarely made much effort to clear the way for the EOD teams in their vehicles.

Frequently the public were downright hostile to the EOD teams and hooligans hurled bricks at the Pigs on their way to answer a call. The need to carry armed escorts indicates the danger from snipers. On my first trip to a bomb in Belfast – it was in a cinema – I was walking down the street with an operator one minute and found myself alone the next. His experienced eye had registered the fact that we were heading straight into an area, overlooked by tall buildings, in which snipers had been known to operate. I like to think his disappearing act was a practical lesson in minor tactics and not simply a low regard for his colonel. Anyway, I took his point. After that I was rarely last into cover at hot-spots.

What is also sometimes forgotten is that although EOD operations are essentially military, they are carried out in Northern Ireland under the civil law, just as the whole presence of the Army in the Province is to support the police in maintaining law and order. If the drivers of Pigs or EOD vehicles of any kind jumped the traffic lights, crossed white lines or broke the speed limits on their way to an incident, they were open to prosecution. There were three cases during my tour. South of Londonderry an EOD driver was booked for speeding and careless driving while *en route* to an incident in Strabane. In Belfast there were two cases, one minor and one which involved a fatal accident. Four people died when a Pig and a car were in a collision at some traffic lights. The Army driver – who was severely shocked by the accident – was charged, but later it was proved that the driver of the car had more than the legal limit of alcohol in his blood.

Traffic lights were always a problem, not only because they slowed down teams urgently required at the scene of an incident, but they made excellent ambush points for terrorists who could open fire at military vehicles either slowing for the amber or halted at the red.

Despite all hindrances, we reckoned we could reach and begin neutralisation of a device within 20 minutes of a team being turned out. As every pound of explosive causes roughly £1,000 of damage, speed is essential. To cut reaction time to minimum a special radio procedure was used. As soon as a device was known to have been laid, the police or an army patrol called their headquarters and the report was passed to 39 Brigade HQ (all Belfast EOD teams came under them unless command had been delegated to some other HQ).

At 39 Brigade HQ, the watchkeeper got onto a secure link to

the EOD duty watchkeeper, beginning his message with 'Felix'. As soon as the magic word was heard the EOD watch-keepr 'hit the panic button'. A bell set the Alpha team moving while a buzzer brought the Bravo boys to the next state of readiness. The ATO destined for the task then went to the watchroom to pick up a form giving the location of the device, the type (where known), the position of the incident control point and – most important – the best route to the bomb. At times of crowds and demonstrations the last detail was critical.

While their leader was picking up his fact sheet, the rest of the team manned the vehicles, which were brought into line with the engines turning over. The time from receipt of 'Felix' to the order 'Go!' was about two minutes.

The Pig was the standard EOD vehicle in town use – a one-ton armoured truck with a six-cylinder Rolls-Royce engine. Pigs can be seen with a variety of modifications: wire mesh over the windscreens, 'bedsteads' welded onto their bumpers, and so forth. As a breed they are mature in years and in their time have mounted guided weapons, been used as troop carriers, radio trucks, armoured ambulances and as towing vehicles for counter-mortar devices. Like their namesakes they are not naturally unclean, but are frequently seen in a scruffy condition after rumbling about Belfast.

The usual formation on a call-out was two vehicles with the ATO in the front pointing the way, but there could be a third. All were in radio contact with each other. One Pig carried Wheelbarrow; if I travelled with a team invariably I found myself sitting on the spare battery box – a most uncomfortable perch.

The sort of ride you had depended on the district you were routed through – in the hard Green areas, the Ardoyne, the Ballymurphy estate, up the Falls, you could expect bricks and stones to be at least part of your worries. The journey was generally the easy bit (though sometimes Pigs have been routed past the bomb in error, to the considerable indignation of the EOD team). What you might find on arrival was anyone's guess.

One morning I set off in search of a team which had been called to a street in North Belfast, and arrived in thick fog just as the cordon was being deployed. As sometimes happened the EOD team were on the other side of the device to me. The

ATO told me by radio that it was an object outside a shop about 100 yards from where I had parked my car. I was standing talking to a subaltern of the 1st Gordon Highlanders trying to get some further information when I heard a sharp report.

'What the devil was that?' he asked.

'Oh, it's just the shotgun on Wheelbarrow firing.'

Then through the fog came a noise I couldn't identify ... a rattle-tattle-rat-tat-tattle noise that was growing louder.

'And what the hell is *that*?' said the Gordon, peering anxiously into the fog.

'I don't know ... never heard it in my life before.'

'Well, you're going to get to know any minute, Colonel, because it's getting nearer.'

We both edged back cautiously. The noise stopped and we waited, ready to duck.

The racket, in fact, had been made by a beer keg, blown over by Wheelbarrow's shotgun, and it had rolled down the road towards us, stopping only 20 yards away. Mercifully it was a hoax or we would have had more than a hangover to nurse.

The real point of the story is the speed at which devices or hoaxes could be dealt with in an urban area, and the need for the speed. The whole beer barrel hoax was dealt with in half an hour and the street was reopened to traffic shortly afterwards. In the country, the need for speed was, in most cases, of minor importance. The number of options open to the terrorists required the emphasis to be placed on a deliberate and careful approach. Any attempt to cut corners could end in tragedy, as was shown in an incident in a country area which had been the scene of much terrorist activity.

A van had been hi-jacked and left on the roadside verge. It was causing no serious obstruction and there was nothing near enough to it to be menaced. I have already described how, when a car was blown up on the border just before Christmas, we left it to 'soak' for some time and eventually dealt with it only because of other circumstances. When we did mount Operation Cupar it was done with all the forces and equipment advisable and, as it was all part of an elaborate plot centred round two milk churn bombs, the cautious approach was more than vindicated.

Much of our success on that occasion was due to the ground force commanders on the spot understanding the philosophy of EOD and the reason for its methods. Both the battalion commander and the company commander concerned had taken a canny Scots view of the situation.

In the case of the abandoned van, the unit involved was a company from an English county battalion. The company commander was responsible for military operations (in conjunction with other sections of the Security Forces) in a pretty tough district and he drove down to look at the van in a civilian vehicle. He then told the ATO attached to his headquarters that he wanted it cleared the following day.

The ATO was a young sergeant who, with more experience, might have argued for more time and more caution. Instead he did as he was told and, after the company commander had flown over the van at first light and reported nothing suspicious, he took his teams to the spot. Within a very short time he had fitted primers underneath it and blown open the doors and windows. Nothing untoward turned up and a physical check of the van revealed nothing. He declared the area to be clear.

It was then decided to move the vehicle from the verge and while a policeman sat at the steering wheel a number of soldiers tried to push it. The van wouldn't budge. A tow rope was produced and fastened from the front of the van to the bar of a Land Rover. It moved off this time and almost immediately disintegrated in a tremendous explosion.

Called out from HQ Northern Ireland I went to the scene by road as fast as I could, having arranged for the DIFS team to be airlifted to the area by helicopter. Senior RUC officers were on the spot when I arrived and gave me an outline of what had happened. The van, or the wreck of it, was a ghastly sight and the kindest thing that can be said is that the policeman who died at the wheel can have known nothing about the exolosion. My sergeant, who had taken a close look at the carnage also, was in a very shaken state and it was not much use questioning him on the spot so I sent him back to his base and retained his No 2.

He told me that the abortive recce by the company commander in the early morning was the sole action taken before the team moved in. No dog team had been used and the Royal Engineers search team had not been called out. By this time the RAOC warrant officer from DIFS had arrived and we checked the area.

The source of the blast was found in the side of the road where, in my estimation, 200 lbs of explosive had been buried. In the hedge we found the remnants of an anti-movement device that we calculated had been placed under the nearside rear wheel. Running up the hill which rose beyond the hedge there was a white flex joined at an observation point on the crest to two 4½ volt batteries. The flex had been there some time and clearly the device had been placed so that it could have been detonated either by someone lying in wait or by the anti-lift mechanism under the rear wheel.

I feel even now that this was a death which could have been avoided if simple precautions had been taken. For a start the recce flight should have been flown by the Royal Air Force and photographs produced. A dog team should have been briefed and, most important, the RE search team should have been alerted.

It is possible, of course, that the white flex would not have shown up from the air, that the dog would not have smelled the explosive, and the Sappers might have missed the flex and the arming mechanism. Any one or even two of these things might have happened but it would have been a very unlucky day if all three operations had proved negative. Even more sadly, there was a piece of towing equipment on the scene, spring loaded and attached to Wheelbarrow, which could have been used to move the van before it was approached.

Once I knew the facts, the ATO sergeant was removed from duty and within two days he was back in England. Subsequently he resigned from the Army. As for the company commander who had instigated the clearance, I considered he had put too much pressure on the ATO to do the job quickly and had shown serious lack of judgment by not calling in the RE search team. I had no hesitation in censuring him in my official report on the incident to the Command Land Forces.

Earlier in this chapter I mentioned the need for officers to understand the philosophy of EOD operations. I should have emphasised that this refers to EOD in Northern Ireland. As far as I am concerned the Province is a special case and the gap between operations there and operations in Great Britain was brought home to me forcefully when I attended an international conference at the Royal Army Ordnance Corps

Headquarters at Deepcut during the period of my tour as CATO.

I was given 45 minutes to cover a period in which my soldiers had been dealing with an average of four bombs a day and had been called out ten times a day. I outlined the major bomb incidents over a period of 12 months and described trends and the introduction of new devices.

The Metropolitan Police took two hours to describe their year, followed by the Home Office Forensic Science Laboratory representatives who explained how they had been meticulous and painstaking in piecing together the evidence provided by the Met Bomb Squad.

I wish that Sergeant B—, who had worked with me in Ballymena, could have been present. He and I had dealt with more devices in one day than they had faced in a year. If we had talked pro rata about 'our' bombs the conference would have lasted weeks if not months.

In saying this I have no wish to denigrate the work of two splendid institutions – I am merely envious of the conditions under which they were able to work and wish to use them to highlight one of my major problems. Unlike the Met and the Home Office, we lacked the time and manpower to evaluate thoroughly the devices in use in Northern Ireland.

There were sufficient EOD teams on the ground to deal with devices, and at no time did we reach the limit of our strength, though on occasions we were stretched, but it was in the backup facilities available to us that we suffered serious deficiencies. During the 14 months I spent in the Province as CATO we dealt with 1,500 bombs and a similar number of hoaxes. Under a perfect system all examples of the remains left after clearance needed to be studied for legal clues, criminal evidence, intelligence information, signs of new trends, origins and many other things. But the means were not in existence.

The RUC provided the highly-trained SOCOs – Scenes of Crime Officers – with whom it was a pleasure to work. At that time, however, there were very few of them (I believe the total reached 14 before I left the Province), and they were required to assist on the spot at all serious crime locations, rape, murder, shootings and bombings. When with an EOD team they were invaluable because they could handle the evidence and sweep the area while the EOD team got on with their next job.

In their absence, however, the team was expected to obtain evidence itself and to this end instruction was given on the pre-ops course. This was expecting a great deal as the legal requirement is for irrefutable evidence that can be linked to the suspected person and has been obtained in the prescribed manner. One of the rules of evidence applying at the time (and I suppose at this moment) was continuity of evidence. In practice this meant that samples of explosives, or containers, or cord, had to be taken at the scene of the incident, packed and sealed in a bag provided for the purpose, and labelled and identified.

Having obtained the evidence, a forensic laboratory report had to be written, listing the samples taken. Samples and reports were then handed into the DIFS lab's explosives section, within 24 hours of the incident.

This detective work obviously added to the labours of the EOD teams, most of whom were based in quarters which were far from conducive to paper work, but the affect on the DIFS people was staggering. In 1976 there was a 40 per cent increase in the workload at DIFS over the previous year and the staff were fully occupied preparing exhibits and giving expert evidence in court. Under this pressure, they were unable to provide the permanent staff needed to give the RUC and ourselves the feedback we required – the collation of statistics and details of the co-relation and attribution of various ingredients.

We required chemical analysis of the sodium chlorate and nitro-benzene used in devices so that the manufacturer and supplier could be identified and this could be done only by comparing forensic data against samples. There were other statistics wanted, too. The backlog of work which had built up at DIFS, and which was being added to every time there was an incident, prevented us making the best use of the evidence obtained.

I got an opportunity to attempt to ease the log-jam when I gave Mr Roy Mason, then Secretary of State for Northern Ireland, an informal briefing at the beginning of 1977. At this I was able to bring him up to date on the types of devices being used and the changes in terrorist tactics, including the use of radio-controlled devices in Belfast and the boobytrap campaign around Magherafelt, an area which Mr Mason had visited more than once. After listening intently for about two hours, he asked me what major changes I would like to see to improve our efforts against the bombers.

Knowing him to be a man as devoted as anyone to beating terrorism, I chanced my arm a little. I asked him to use his influence to enable me to make on the spot visits to the scenes of incidents or arms finds in the Republic, and I also asked him to increase the establishment of the explosives section of DIFS by two higher scientific officers and a scientific officer. He promised to do what he could for me and I duly presented him with a paper containing my case.

The Secretary of State was as good as his word. He did try to make an arrangement whereby I could visit the South on appropriate occasions, but this was never a starter really and did not come off. As far as my other request was concerned, he set the wheels in motion through the proper channels and a series of meetings were held, chaired by the head of DIFS, and attended by the head of the CID, representatives of the Northern Ireland Office, and myself. It seemed like a good idea at the time ... but there was a question about cost. Although in the overall picture of what the troubles were costing, the sums were insignificant, that strange civil service chemistry came into play: the smaller the sum involved, the greater the argument about it.

How long, it was asked, would the terrorist campaign last? When would the present high level of bombing decrease? And just supposing these three scientists were established and suddenly peace bloomed? What would they do with three surplus scientists?

The whole point of having the extra staff, of course, was to hasten the advent of better times, but that did not seem to be understood. As to visualising the end of the troubles, anyone's guess was as good as mine, but from the bombing point of view the reason I wanted better back-up was that things were getting busier – there was no slump in the terrorist trade.

Ah, but it was not the job of DIFS to engage in intelligence gathering. They sat back as if to say, 'Got you there!'

There is, in fact, a narrow dividing-line between intelligence gathering and forensic investigation. But if it comes to strict divisions, it was not really the task of the Army to do a number of jobs. Besides, DIFS had a department specifically created to work on explosives – it was simply a question of enabling it to function efficiently. Still the grumbles went on. The police, government research stations and the Home Office were all dragged into the search for some other means of giving the service required – a service which was sadly overworked at that time.

I became convinced that bureaucracy was more concerned with observing the rules concerning employment of public servants than they were with convicting terrorists who obeyed no rules at all. It was a ludicrous situation to me, although I admit to being biased. After eight years of bombing we were faced with a backlog of work. Already I was sending certain technical problems to research stations for evaluation because of the lack of proper equipment to study electronic devices in detail and the Army's own intelligence organisation stopped far short of major assistance in the field of explosive devices.

Fortunately the long-suffering RUC supported me in my arguments and we gained approval for the additional staff. The increased manpower proved to be invaluable. Even though the level of bombing dropped later it meant that the difficult processes connected with forensic examination of explosives was being done thoroughly and, equally important, comprehensive records were kept and proper deductions made. Our own forensic work went on as usual, and it was encouraging for the men on the ground to know that the time spent in filling up forms, labelling bags and delivering to DIFS was not being wasted.

Naturally, because of the dangers in handling HME (homemade explosive) only the smallest amounts were returned to the lab, the rest being destroyed under controlled conditions. Rare mistakes were made and assembled devices were sent to the lab. The culprits, protest as they might that 'it wouldn't go off in a million years', were dealt with under the usual military disciplinary code. I did not want to take the slightest risk of DIFS being blown up.

CHAPTER TEN

The Booby Trap Campaign

The purpose of all acts of terror is to terrify, though attempts may be used to justify and even glamorise the most indiscriminate and appalling attacks. The aim of terrorists is to seize control of their objective and exercise power, either directly or through men and women who are more acceptable to society but are mere tools, however self-deluded, also in the grip of terror.

Terrorists discard all attempts to reason with opponents the moment they try to convince someone of the justice of their cause by killing them. To work off their frustrations, such terrorists are likely to invent more excesses even though this affects their own and potential supporters. Occasionally the *Irish News*, a Belfast daily newspaper with broadly Republican sympathies, carried 'In Memoriam' notices, reflecting the ferocity with which the Official and Provisional wings of the IRA sought to resolve their differences a few years after the terrorism rose to the surface. As it was not unknown for an 'In Memoriam' notice to appear before a victim had been executed, it was always as well to scan the column. A reader might find his own name there.

Fainthearted Republican sympathisers know that among punishments meted out is the piercing of an offender's knee cap with a household electric drill, generally said to be a Black and Decker.

While I was in Northern Ireland I heard of more than 50 cases of men who had been crippled by a shot in the knee with an ordinary pistol to persuade them to put their backs into the fight to unite the land of Saints and Scholars. The use of tar and feathers on girls who might have looked too kindly on a stranger seems almost soft by comparision.

Perhaps the escalation of the indiscriminate bombing and

murder campaign of 1976 was due to some megalomaniac's wish to discipline the whole of the contrary Six Counties for not collapsing under the strain. Then a mind, warped in a different direction, must have realised that perhaps it was counter-productive to maim and butcher civilians, of whom 245 were listed as dead at the end of that bloody year, the worst for the public since 1972 and, as it turned out, the second worst for them during the period 1967–1978. In August, the death of the three McGuire children, run down in Belfast by a fleeing IRA gunman after he was shot, caused a wave of revulsion and the formation of a Peace Movement. Personally I doubt whether this was instrumental in inducing a change of tactics, because the unbalanced minds of the terrorists can make out a case that even these children died for their cause, but change they did. The emphais was switched from general to specific targets.

The terrorist option was fairly wide – the Provisionals were (and are) seeking to undermine established institutions, disrupt industry and services, weaken confidence in the Security Forces (and cause the SF to doubt themselves), and obtain the maximum propaganda effect. In the event, they intensified the attack on a section of the Security Forces in the shape of the Royal Ulster Constabulary and the Ulster Defence Regiment. Harsh though the consequences were, the choice of target reflects the effectiveness of the two forces.

At that time the strength of the Security Forces in Northern Ireland was, in round figures, 14,000 soldiers either on 18-month garrison duty or four-month emergency tours; 6 000 RUC and RUC Reserve; plus more than 8,000 UDR personnel.

From a propaganda point of view the RUC and UDR were likely to be fruitful fields, at least as far as preaching to the converted was concerned and for those who did not know the background to the situation. Both forces could be bracketed with the 'B' Specials, whose vilification and denigration is an historic plank of Republican policy.

The 'B' Specials were a volunteer part-time force formed between the wars to support the RUC – originally there were 'A', 'B' and 'C' Specials, with different rates of pay and obligations. Like the RUC they were armed, holding their weapons at home, and they operated in their own backyards,

so to speak, being strongest in country districts. The organisation was of a para-military nature, under the leadership of its
own commandants answerable to and directed by the police
authorities.

The complaint against them was that they were sectarian
(they were solidly Protestant for all practical purposes), un-
disciplined and apt to take the law in their own hands, or at
least to bend the rules. Occasionally they were accused of bias
and brutal behaviour. In their favour it was argued that, in
general, they had been effective over a long period with little
official financial backing, that nearly all the charges levelled
against them were exaggerated grossly and that in any case a
smear campaign on a large scale was being conducted against
them. Finally, they were necessary as no other body existed to
do the job they were required to do.

I see that in *A Place Apart* by the respected Irish writer
Dervla Murphy a glossary is printed in which the 'B' Specials
are referred to as 'an armed auxiliary police force, once the
para-military wing of the Orange Order but now disbanded by
order of the British Government'. Although the text qualifies
this with the statement that they were the 'unofficial' wing, the
effect is the same – wittingly or unwittingly, the propaganda is
perpetuated. The Specials, or the hated 'B' Specials as certain
elements of the media on both sides of St George's Channel
call them in parrot fashion, were disbanded in 1970 after the
publication of the Hunt Report and at the same time the
regular RUC were disarmed. Many people I met in Northern
Ireland, not all of them Protestants, regarded the two events as
the greatest propaganda coup of the campaign. In one stroke it
demoralised the professional policemen who had been taking
the strain at a time of great unrest, while simultaneously
removing an established reserve, albeit an imperfect one,
which was a major source of intelligence and information.

In order to fill the gap they had created, the government
then formed the Ulster Defence Regiment, to be recruited to
serve in Northern Ireland as an integral part of the British
Army, commanded by officers from other regiments and with
a cadre of officers and NCOs from the rest of the Army – training majors, quartermasters, regimental sergeant-majors and
permanent staff instructors. The men were recruited locally,
did their duty in uniform two or three nights a week and/or at

weekends, and were subject to military law while on duty. (Officers are subject to military law at all times.) As part of its charter, the UDR was required to be non-sectarian and existed to protect the state and the border from armed attack.

At the outset the Regiment included a large proportion of former 'B' Specials and most recruits were Protestant. Significantly, however, Catholic membership reached nearly 20 per cent of the total. Naturally, this did not suit the IRA and a campaign of intimidation was launched to drive out the Catholics, a move that intensified after the introduction of internment in 1971.

Both the RUC and the UDR went through hard times during the early years of the troubles, the former having to rebuild confidence and the latter having to establish themselves from scratch. In the case of the UDR this meant overcoming the doubts, suspicions and criticisms of other Army units serving in the Province. I even had an ATO who refused to clear a device because the cordon was provided by UDR and not the local regular battalions.

By 1976, however, despite a variety of setbacks, it was clear that both organisations had become vital and effective parts of the Security Forces. The police, who had long since been rearmed, were operating once more in some hard Green areas and the UDR, though its Catholic membership had fallen to two per cent (the ex-'B' Special element had decreased significantly also), was becoming increasingly professional. Its turn-out for duty during the Loyalist workers' strike in May 1977 was almost one hundred per cent, thus completely confounding the Jonahs who forecast the regiment would not act against a Protestant-dominated organisation. As the UDR PROs (both English and both Catholic) used to point out in the mess at Lisburn, 'The Republicans and the Loyalists both condemn our boys so they *must* be doing their job properly.'

I have given a brief outline of the position of the RUC and the UDR to explain why the terrorists should see them as prime objectives. Here were two indigenous organisations, improving in strength and efficiency, patently helping to stabilise the community and to protect it from groups who were trying to destroy it. Clearly it was in the interests of all terrorists, including Loyalist para-militaries, to weaken them and to deter men and women (both forces have a female element) from joining. The IRA went about the task by assassination – using either gunmen or booby traps.

Personnel serving with the RUC or the UDR are particularly vulnerable because of the exposed nature of their duties. The police are just as much engaged in routine tasks, traffic control, crime, assaults and dealing with drunks, as with terrorism. The uniformed branch, in green so dark as to be almost black, with a little splash of scarlet behind the harp badge, and in their light green shirts, are a familiar sight. They were not operating freely in all areas during my tour but in enough places to expose them to the touts working for would-be gunmen. Little wonder many of their duties required them to wear body armour and carry arms.

The UDR soldiers were also plainly visible, operating, as they did, vehicle checkpoints in the city and in the country. During periods of heightened tension, when all or part of the eleven battlions of the regiment were mobilised for full-time duty, they took over border crossing points in Londonderry and at other sensitive points. It was not difficult for subversives, therefore, to get a look at them.

It was even more likely that the neighbours of serving members of both organisations would know whether or not they were serving. A UDR man could take all the precautions recommended – and all were briefed on personal security – but his wife might still hang out his combat uniform on the clothes line on wash day. This happened more than once. In the more remote country areas, it was even less likely that a man's affiliations could be kept secret.

It was all the more remarkable therefore that, despite the known perils, so many folk came forward to join these branches of the Security Forces. For them, unlike the soldier on a four-month or 18-month tour, there was no question of looking forward to the end of it and a new posting to, say, Germany. They were there for the duration ... for life, in fact, because even those men and women who left the RUC or UDR after giving good service knew they were marked down as far as the opposition was concerned. Yet while I was serving in Northern Ireland there were plenty of 'part-time' soldiers in the UDR who had been on duty two or three evenings a week for seven years – and who may be at it still for all I know. A number of them were veterans of World War II.

The problem that faced such people was that whatever steps they might take for their personal safety and that of their families, however much they might vary the routes they travelled, and the times they left the house, no matter what drills they had for answering the door or simply closing the curtains (for those who don't know you never put the light on until after they're drawn), the fact remains there were three places they would eventually be found. In the case of the UDR, the danger spots were their homes, their work and the UDR Centre (for a police reservist it would be the police station).

There had been a flurry of activity against the UDR just before I arrived, particularly around Magherafelt where 'F' Company of the UDR's 5th Battalion was based. A private in the regiment was murdered by gunmen while at work near Toom Bridge in April and the following day a sergeant was riddled by terrorists with a Thompson sub-machine gun at Gulladuff in the same region. The sergeant was going about his business as a postman when his van was ambushed.

In July another member of the same company was killed. Private Robert Scott, 26, was well known for his kindly disposition. While working on his father's farm, he found time occasionally to help out an 86-year-old widow who lived in a cottage on the land. On the morning of 30 July Scott decided to look in on the old lady. He was going to chop some firewood and make her a cup of tea. Getting down from his tractor to open a gate he triggered off an explosion which killed him. When I investigated the incident I found a length of fishing line and a drawing pin attached to the bottom of the gate. The device itself had been cleverly concealed.

As summer drew into autumn the pace got hotter. A fair-haired young man walked into a grocery in Armagh City and asked for a pound of cooked ham. As the shop manager turned to lift it from a shelf he got a bullet in the back and the UDR's 2nd Battalion had lost an experienced lieutenant. In Tyrone, a postman delivered a letter to a lonely farmhouse and was shot down as he walked away. Terrorists had sent the letter, then gone to the farm and held the occupants prisoner until it was delivered. The dead man was a corporal in the 8th UDR which had its headquarters at Dungannon.

As if not to be outdone, bombers began to show considerable ingenuity – emphasising the fact that in combating terrorism it is essential for the security forces to use their imagination in protective operations.

A company of the much-battered 5th UDR was parading at the ranges at Portballintrae to practise musketry. They fired a number of rounds at 100 metres and then moved back to the 200 metre firing point. As they were taking up position, a soldier who had been distributing ammunition stepped into the shingle that filled one of the positions and was severely injured in the explosion that followed.

The range had been in use the previous day. It was being covered against possible attack by another UDR unit, and was under reasonable guard. I flew up from Lisburn to find out what I could and to join Sergeant B— who had been called in from Magherafelt (it was only a week before the fire raid on Ballymena).

The RE search adviser was on the spot when I arrived, having come all the way from Ballykelly, the old airfield near Londonderry. He was none too happy and I soon found out why. The officer in charge of the operation, a major from a county regiment, seemed to have some strange ideas. When I told him that I was going to the firing point to collect evidence – I had brought a warrant officer from DIFS and a scientific officer with me – he refused me permission. He told me he had decided to do a full clearance operation the following day and I would have to wait. I couldn't believe my ears.

Had the device been discovered and isolated there would have been some merit in delaying, calling in helicopters, staking the place out, and so forth. But the place had been staked out during the weekend, the UDR had walked all over it, and the casualties had been cleared from the firing point.

In theory I could have pulled rank, but he was in command of the operation and such an action on my part might have led to bad feeling at a later date. I decided to humour him.

'What size boots do you take?' I asked.

He looked bemused.

'Size 10, sir. Why?'

'Well, I take size 8 – so if you walk in front of me, and we follow the same path as everyone else, we should be all right, I reckon.'

The proposition was well and truly considered before the major agreed and he actually did walk in front of me while I

124

followed in the track of his size 10s to collect what evidence I could.

A further operation was carried out to clear the area the following day, and the evidence finally pointed to 70 lbs of home-made explosive having been buried in a milk churn at the firing point some time previously. The terrorists had probably slipped back during Saturday night – Sunday morning and installed a pressure plate consisting mainly of two boot polish tins which they buried six inches under the shale and chippings. No-one had spotted any difference in the gravel – and there could have been little or no chance of anyone other than a dog sniffing out the booby trap.

Digging in explosives where they could be primed when required was a standard method of operation by terrorists, especially on the border. A perfect example was the discovery of four oil drums filled with ANFO dug into a bank alongside a road, near the South Armagh border, complete with booster, and linked with Cordtex. They had been there for a few weeks before they were spotted — heavy rain caused some subsidence which exposed the tip of one drum just below road level. All the device needed was an initiation system and it could have been activated by a cable or a radio-controlled system at any time by terrorists popping across from the Republic which was only 200 yards away.

Today, of course, it takes only a few lines to describe the device – but it took WO2 S— the best part of two days to neutralise it and discern its nasty nature. Now we know what it was, we can apply hindsight to the situation and consider it from all angles. This benefit, however, was denied to S— whose logic had to be applied in the light of experience only. After the aerial survey, the sniffer dog, the search team and sweeps to locate the possibility of rado-controlled initiation, it was still a matter of a man putting on a bomb suit and attaching the inevitable hook and line to each drum. It was up to him to decide in each case how long he would leave the thing to 'soak', and, in fact, the laborious but necessary precautions were drawn out until the light began to fail and he wisely decided to carry on the next morning.

In the end this painstaking warrant officer had the satisfaction of destroying 27 lbs of booster on the spot and of spoiling 700 lbs of home-made explosive with ditch water. The latter, he revealed in his report, was beginning to crystallise, which meant it was becoming highly dangerous.

Only two days earlier, S— had neutralised a milk churn containing 70 lbs of ANFO which had been placed outside the Derrylin telephone exchange. He had gone through the preliminaries and was in a builder's yard opposite the device when someone moved a brick from a pile beside a shed. A black object with a wire attached to it came to light and S— and his team had to beat a hasty retreat. A shotgun blast from Wheelbarrow showed this to be simply a place where the builder kept the key to the shed – but it illustrates the surprises which face operators and the distractions which occur. The explosive in the churn, by the way, he cleared with a hose attached to Wheelbarrow and operated remotely.

The Derrylin bomb was, of course, a simple straightforward attack with a timing device. The massive drums S— had dealt with in the embankment were placed in ambush and it may be as well to dwell here on the difference between the ambush bomb and the equally unpleasant booby trap device.

The first remains under control of the terrorist right up to the moment of detonation. The booby trap, on the other hand, is initiated by the victim.

There are two types of ambush bomb – namely, remote-controlled (sometimes described as command detonated), and radio-controlled.

The remote-controlled bombs could weigh anything from a so-called Claymore of about ½ lb to hundreds of pounds of explosive packed into a culvert. Control wires led from the explosive to a battery pack where a terrorist waited to operate the switch which shorted out the circuit. The distance from the terrorist to the mine depended on a variety of things such as the power of the batteries and the wire resistance.

Most people will be familar with the culvert bombs' description but the Claymores may not be so well known. One, which we nicknamed the Shotgun Mine, consisted of a foot-long tube, two inches in diameter, welded to a base plate (like a crude mortar). A carrying handle was welded onto the side. With a sugar chlorate charge as a propellant for 2lbs of assorted ballbearings, the idea was to dig in several on each side of a road so that when they were set off by their controller – the wires activated a standard flash bulb igniter – the contents scythed over a wide area for a range of 100 metres.

The Bucket Claymore was a bigger device. The container was a standard galvanised bucket with a hole in the bottom for the detonator, around which was packed 2lbs of commercial explosive. Some 4lbs of CO-OP went on top of this and the whole lot was topped up with 15 lbs of 'shipyard confetti' – nuts, bolts, screws, door fittings. These devices are designed to be buried into the side of the bank of a road which the Security Forces patrol and are very powerful – the blast damage alone of 6lbs of this explosive will extend over a very wide area.

In country areas the troops on the ground were well aware of the menace of remote-controlled or command-detonated bombs. If wires were spotted the area would be cleared immediately, cordons set up and set drill put into operation perhaps involving search teams, dogs, EOD, and aerial reconnaissance. Though it is the most natural thing in the world for a soldier to want to follow a wire to see what is at the end of it, this temptation was usually resisted. This caution was also exhibited in Londonderry where the compactness of the city and the nearness of open spaces made troops aware of the threat in both urban and country areas. The same was not always the case in Belfast where certain of the *roulement** battalions seemed to take a delight in getting on with the job before the arrival of the EOD team. In doing so they exposed themselves not only to unnecessary risks but were likely to destroy valuable forensic and intelligence evidence. Where EOD teams did get to the scene they could, for example, tell whether or not a circuit was live. Provided they knew which was the business end they could destroy it *in situ* without going near it. And they could also neutralise it.

Ironically the increased use of the ambush and booby trap bomb was due to the success of the Security Forces north of the border and the Southern Irish authorities' increased pressure on plants producing explosives in the Republic. In 1976 the amount of explosive known to have been used by the IRA and associated organisations was only half of that in 1973 (though it still amounted to 25 tons). This was due mainly to the fall in the number of cars used as containers for large amounts of explosives, and to the increase in the use of blast and incendiary devices. In proportion, however, the percentage of

* The battalions on a four month emergency tour.

ambush bombs and booby traps peaked in 1976–77 and this enhanced interest in direct attack on members of the Security Forces continued until Easter 1977, when there was a general lull in bombing activity. The IRA continually changed the emphasis on the type of bombing carried out because of either the success of the Security Forces against them or the fact that a particular tactic had outlived its usefulness. Probably a combination of both these factors led to the sudden cessation of the intense booby trap campaign in the spring of 1977.

Finds in 1976 had provided evidence that the IRA were quite capable of producing sophisticated devices. Micro switches, infra-red, light sensitive, and photo-electric elements were all likely to be used against us. If the terrorists did not concentrate on producing these it may have been because they argued, why go to the trouble of wiring a complicated circuit when a simple mechanical switch will do the job equally well?

Perhaps more difficult than making the booby trap device is the task of selecting the target and pin-pointing his movements. His habits have to be watched carefully, the colour and registration number of his car noted, and all routine behaviour recorded. The Provos sent out children to check car numbers on an estate in which officers and their families lived.

The man who observes a set pattern is instantly more vulnerable than the unpredictable individual who varies, say, simply his time of leaving for work and the route to it. In my time there was no such thing as routine convoys of military vehicles, for example. The roads used were changed and so were the timings, the constitution and the number of vehicles in the convoy. For individuals it was more difficult. The terrorist will persevere whatever precautions people take. In their determination to make a point they are well aware that even if their main target escapes, members of his family, his wife or children, will provide subsidiary victims.

Under the normal threat in Northern Ireland it is unwise to neglect basic security precautions. In the country areas of Londonderry no one could consider himself safe in the winter of 1976–77.

No doubt, over the centuries, there have been times when men could pause and admire the majestic sweep of the Sperrins or fish lazily in the pleasant waters of the River Roe, but it was salmon line and a different game that held the imagination in the later months of 1976 and early 1977. Not for the first time in its history was the area particularly inhospitable to the forces of law and order and their supporters. Truly there is nothing new under the sun....

The legendary O'Neills of Ulster, earls of Tyrone, dominated the region in the seventeenth century and in 1641 they were prominent in the insurrection which saw Moneymore overrun and Bellaghy (then called Vintner's Town) and Magherafelt sacked and burned. Nine years later another of the O'Neill faction, Bishop Emer MacMahon, conducted a skilful little campaign against Cromwell's forces, seizing Toome, strategically placed where the Bann joins Lough Neagh. He was thus able to threaten a number of objectives before his enemies could join forces to attack him. This warrior priest, the Bishop of Clogher and a former Vicar-General, then swooped on Dungiven at the foot of the Glenshane Pass and summoned the fifty-strong garrison of the earth fort to surrencer. A furious fight followed their defiance and finally all were put to the sword. Neatly dodging his pursuers by slipping across the Foyle at Strabane, the warring bishop then made a serious error. Against the advice of his professional officers, he stopped to fight, letting his men leave a strong position to attack the enemy. They were caught on bad ground and routed within an hour. Trapped near Enniskillen after a relentless pursuit, McMahon was taken captive to Londonderry where he was summarily hanged.

Had he or his followers been able to come back and retrace their steps during the period of this narative they might have concluded that not a great deal had changed. Toome, with its extensive eel fishery, was still of strategic importance and bore the scars of recent attack by more than one massive car bomb and sundry other ugly incidents. From the slits in the sangars * of the wired-in RUC station young soldiers of the Royal Hampshire Regiment watched attentively as the hurrying traffic slowed to roll carefully over the 'sleeping policemen' checking the pace of vehicles using the causeway over the Bann. Walls in once-burned Moneymore were marked with the bullet splashes where a patrol had come under fire. On the road to the west lay the devastated main street of Castle-

* Fortified look-out posts.

dawson – half a dozen fine old cottages shattered in one terrific blast from a 500 lb car bomb. One day the shade of the savage old soldier priest could have witnessed the funeral of a UDR part-time soldier cut down by gunmen as he stood with his son at a bus stop in Bellaghy – the next a stouthearted young farmer, also a UDR soldier, hurling himself through his own front window to escape the automatic fire of terrorists who had waited to ambush him on his return from duty in the early hours of the morning. Fifty holes in the walls of the farmhouse testified to the intensity of the fire and the luck and courage of the man who dodged it.

In these contested lands, where three hundred years ago the fierce Scots settlers pushed west across acres once the property of the ancient Irish aristocracy, the atmosphere was sometimes even more sombre than that on the Armagh border. In South Derry the past, with its legends of enduring hatreds, seemed nearer. The shade of Bishop MacMahon would have noted the growing list of names on the Roll of Honour in the UDR Centre at Magherafelt and the bustle as men of the Special Patrol Group moved in to join search operations. In the small towns a few miles around other shades were watchful too, shades more concrete than Bishop MacMahon's.

The winter boobytrap campaign was a considered and well orchestrated attack. This was not the usual run-of-the-mill affair, in which a bomber was lent to a PIRA unit to carry out specific tasks. On this occasion, in my opinion, a trained team was sent across the border for a prescribed period, at the end of which it was withdrawn. I had always considered that the best boobytrap devices were produced in the Strabane area – the clock used in the last tanker I cleared in Belfast was of a type only used in Strabane up to then – and that dolorous region was within convenient range of the target area in South Derry and the shores of Loch Neagh.

The first strike was made just before Christmas. A member of the RUC was setting out for work in the usual way and his wife was standing in the doorway of their bungalow waving goodbye. He got into the car and drove ten yards down the drive when there was a shattering explosion. The force of the blast tore off the man's legs below the knee and hurled the torso into the back of the car. In a few split seconds on a day which had appeared to be normal and even humdrum, a man had been killed, his wife scarred for life and a family's dreams and ambitions obliterated. Two pounds of explosive had been secured by an intruder under the right-hand wheel arch. The timer was the old clothes peg device with the jaws held open by a wooden dowel. A length of fishing line from the dowel had been led through the wheel and hooked onto the tyre valve. As the wheel revolved the line tightened and pulled out the dowel. Contact was made and the explosion followed.

Within a few hours a similar device killed another RUC man, and in the weeks that followed the Provos did everything with fishing line. They attached it to doors, across doorways, between radiator grills and to garage walls, around prop shafts, on tyres, through tyres, in fact anywhere where it was likely to be overlooked.

At Garvagh, north of Magherafelt, an RUC reservist who had parked his car outside his home at midnight was having breakfast when he noticed that a line, almost too thin to be visible, was hanging from the radiator grill of his car. Sensibly he phoned the police and the local EOD section was alerted. The sergeant ATO, from Magherafelt, came out and was joined by S—, OC 321 EOD. A length of salmon line hung down with a loop resting on the bumper. It was the sort of thing the inquisitive might be tempted to pull but after various other methods had been tried, the bonnet was blown up with a controlled explosion revealing an unpleasant looking paper parcel on the clutch housing. The ATO then donned his bomb suit and attached a hook and line to the device which was pulled clear. It consisted of more than 2½ lbs of commercial explosive and was successfully disrupted.

The following day another reservist in the RUC popped out of the back door to check the area and open the gate into his drive. On his way back he noticed three drawing pins in the front door with a length of fishing line stretching from them. Once again the Magherafelt sergeant was called out and went through a series of drills to deal with 5lbs of explosive wrapped in a piece of paper cut from an old cattle feed bag and placed on the target's doorstep. Had the door been opened the front of the house would have been blown in.

'These things come in threes,' is an old saying of which the bomber himself was clearly aware. He attacked another target the following night and in the morning a regular sergeant of

131

the RUC spotted the tell-tale fishing line when he left by the back door to clear the surrounds of his home at Maghera, north of Magherafelt. The line was used as a trip wire in this case, stretched from the device in a plastic ice-cream container to a decorative flower pot on the other side of the door. A person's leg pushing through or catching the cord would have pulled the clothes peg initiator and set off 5 lbs of commercial explosive.

There was at that time a standard form which operators had to fill in after a clearance. It contained among other things a square with the simple word 'Target'. In Sergeant K—'s report this square contains the laconic description 'Regular RUC Sergt/ATO'. Instead of ATO he could well have written 'myself'. As the terrorist had guessed, the man who had been successful in disrupting his devices on the two previous days would be tasked to deal with the device on the third. Fortunately K— was a shrewd soldier. By judicious use of Wheelbarrow and a selection of disruptive devices, plus the application of the hook and line, he neutralised not only the explosive in the ice-cream container but an equally deadly device which had been hidden in the flower pot. Had this been even lifted it would have killed the handler immediately. Instead, thanks to Sergant K—, we got an excellent sketch of a simple tilt switch.

The failure of the grisly salmon fisherman to catch his human prey was due to a display of commonsense, observation of the rules and a lot of courage. Because of it K— was able to fill in for the third day in succession another important box on the report form, the one marked 'Reason why device failed to function'. He wrote simply 'EOD action'.

A few days later the bomber tried another tack and fastened his deadly line to the back door of a policeman's house. The bomb went off as the handle was turned. No one was killed but the officer and his wife suffered minor injuries.

From east of Magherafelt a good road leads over the Glenshane Pass through tranquil fields and picnic areas with permanent tables and trestles set out for sunny days and peaceful motorists. Here and there this road has been widened, some say to make it more difficult for terrorists to hi-jack lorries. (The more cynical claim it is so that they can seize bigger ones.) Over the top you drive, with outcrops of rock and quarries on the right. On the left the landscape falls away, then rises green and dotted with sheep all the way to the foothills of the Sperrins. A sudden turn, a few houses and, as you brake, you find you are in the broad main street of the unforgiving and perhaps unforgiven town of Dungiven, where Cromwell's men died so gamely in 1650.

A UDR officer once told me his father in the Royal Irish Constabulary had been fired on and wounded five times in all at Dungiven, and it did not surprise me at all. In my day first the Worcestershire and Sherwood Foresters and later the Royal Hampshire Regiment manned the cramped RUC station and its stifling box-like ops room lined with photographs. The faces are of wanted men and women, who stare blankly at the weary subalterns and gritty, caustic corporals who address known villains by their Christian names. 'It worries them you know. . . .'

Wire netting festoons the front of this outpost and steel shutters protect the windows and keep in the smell of cooking. Entry and exit through the wind-rattled corrugated iron gates is quick and soldiers know they must be on the alert from the time they arrive until they are relieved for a spell. They are natural target material for brooding bombers.

One of these displayed a particularly macabre ingenuity after a member of a foot patrol of the Hampshires was killed by a shot aimed from a derelict building at almost pointblank range. The assassin used a shotgun.

Though the house was cordoned off immediately, the killer escaped. A murderer may find friends in Dungiven. The Hampshires searched the place, noted carefully the scanty soiled contents, but wisely touched nothing. Instead they reported that they had seen only a shotgun cartridge case in a downstairs room. A follow-up patrol showed the same caution and the Magherafelt EOD team was tasked.

To all intents and purposes there was little to fear – that is if you were not a professional. Fortunately the operator was. The shotgun cartridge lay beside a copy of *The Sun* open to display the charms of the nude on Page Three. The operator was not distracted. He felt carefully round the cartridge and discovered that a wire was attached to it. After observing necessary drills the wire was traced under the floorboards and ended up at a large bomb under the doorway. Had the cartridge been picked up the whole building would have gone up and with it the men of the infantry patrol in and around it.

Other devices were tried and some caught or nearly caught the unwary. In Strabane a boy picked up an empty 7.62 rifle magazine and lost his hand when the bomb it was attached to exploded. A man about to get his car out of the garage noticed the door of the building was held shut by a rock which had not been there the night before. The EOD operator discovered a cleverly disguised anti-disturbance device. Near Kinawley a police reservist was shot and wounded, and in the follow-up at the firing point a detective picked up a clip from an American Garand rifle complete with eight rounds of ammunition. Fortunately he mentioned that there was something suspicious in the grass underneath it and the NCO in charge of the unit search team of the 9th/12th Royal Lancers stopped him from picking that up too. A very unpleasant bomb was eventually unearthed under a patch of bramble bushes. It was fitted with an anti-lift device and a planned operation was carried out to neutralise it.

The presence of a bread-wrapper and a sardine tin label resulted in the unusual query in the operator's report: 'Sardine sandwiches while they worked?'

One result of this increased boobytrap offensive was that anything that looked suspicious was treated with respect and empty beer tins remained in the gutter instead of being kicked.

Down in Magherafelt a new piece of home-made equipment was added to the EOD section's 'golf bag' of kit – a garden rake fitted to an extended handle. Mounted on Wheelbarrow it proved invaluable for turning over suspicious objects, such as empty cartridge cases at the scenes of shootings, and it saved the life of at least one operator.

Use of the fishing line and clothes peg switch in bombs placed by the IRA had an ugly side effect when Loyalist terrorists adopted the same technique. Booby traps had been fitted into three cars during December and January, the devices being wired to the ignition system in cars with bonnets which could be opened from the outside. Then, in a TV programme someone went into the details of using fishing line. Within the next few weeks fishing line was used in booby traps in six vehicles, the cord being attached to wheels and prop shafts. In one, on 25 March, one man was killed and five injured when a

Ford Transit van was blown up.

Perhaps the most unpalatable side of the boobytrapping campaign was the problem that arose when the IRA murdered someone south of the border after a kidnapping and then dumped the body back in Northern Ireland. It had to be assumed automatically in these cases that the corpse might be wired up to a bomb and dealt with accordingly. It was with some distaste therefore that I heard the news that a body, believed to be that of Captain Robert Nairac, the Grenadier Guards offier captured by the IRA on the border and then murdered in the Republic, had been discovered in a quarry at Jonesborough.

Jonesborough is as evil as Crossmaglen – an enclave of the North surrounded on three sides by the South and notorious as an area for terrorist activity. If the IRA wanted to try anything they could not have chosen a more suitable place.

Though I expressed surprise that the killers had decided to return the body, as I did not think they would wish to provide evidence of their own brutality, a study of aerial photographs of the area seemed to prove me wrong. One of the experts who interpret air photographs pointed out that in a ditch opposite one of two quarries was a large plastic bag of the type sometimes used for carrying carpets. The shading on the photograph indicated that a body was in the bag – you could imagine a head, torso and knees.

A gaggle of Press men and sightseers gathered in the Republic to watch as search teams went about the task of clearing the quarries. All day they waited while we worked, but their time was wasted. All we did was recover an empty plastic bag – the air beneath it had given it the appearance of concealing a body.

I felt a sense of relief when the truth was known. Although I was anxious that Robert Nairac's body should be found I did not wish to have the responsiblity of perhaps having to violate it to ensure that it was not boobytrapped.

In combating booby traps the increased use of radio-controlled devices presented a special challenge to the EOD sections. This was because I insisted that every effort had to be made to secure a device intact once it had been detected.

As its name implies the radio-controlled bomb is detonated by radio signal instead of by a current passed down a command wire. This means the terrorist can take up a position a considerable distance from a concealed device and can set it off at will, as long as he has line of sight between the transmitting and receiving aerials. This makes such weapons ideal for ambushes such as occurred at Warren Point in 1979 when a detachment of the Parachute Regiment was blown up as their vehices passed a concealed device and a patrol which raced to the scene suffered a similar fate. The incidents took place in clear view of the Republic across Carlingford Lough and the initiator of the outrage could have been hidden anywhere along the opposite bank. A similar bomb inflicted heavy casualties on a vehicle patrol in Tyrone soon afterwards.

The use of RC devices had begun before my tour but was being stepped up while I was in Northern Ireland. They made their first appearance in Belfast itself while I was there, causing me to express misgivings to Commander Jimmy Neville, at that time head of the Anti-Terrorist Squad at Scotland Yard.

All in all, 14 RC devices were used – as far as we were aware – during my tour. They were sophisticated and at least one, which was dealt with near the Kilnasaggart Bridge, in South Armagh, contained a self-destruct mechanism. This was prepared so that if the bomber missed his target for some reason – perhaps a malfunction – the device would automatically destroy itself and the evidence that went with it.

One of the last tasks I carried out involved a device at Dungannon and I was called in because someone had reported that there appeared to be a problem – a misunderstanding between the EOD operator and the Life Guards who were patrolling the area.

As it turned out there was no difference of opinion among the men on the ground, simply a flaw in communications. There was a problem however, and I was glad that I had been called out. The device had been discovered by a patrol and lay in the corner of a field. Detection equipment had identified it as a radio-controlled device. Normally this would have been dealt with according to the book but there was an additional snag – the field was overlooked on three sides by houses. And a number of these were known to contain Republican sympathisers.

I found out from the escort that certain of the houses were

occupied by relatives of men 'on the blanket'* in the Maze prison and this gave an added dimension to the problem.

'I assume you've evacuated these houses,' I said, turning to the RUC man on the spot.

He looked at me as though I had taken leave of my senses.

'Not at all, sir!' he replied indignantly. 'If we tried to do that there would be a riot.'

'Then what have you done?'

He leaned forward confidentially.

'We've had a chat with them, Colonel.'

'And?'

'They've all agreed to move into the back rooms of the houses where they won't be seeing a thing.'

Clearly a man of a calm and serene disposition, he stepped back and beamed, waiting for the next move.

'OK, you've heard what the police have said. Now listen! Staff and I are going to take a look down the road. If you see any movement in any of those windows, shoot. We'll answer questions later.'

The staff sergeant was the EOD operator with the Armagh team and he had explained already that because of the awkward location of the device he would not be able to use Wheelbarrow. He intended to make a manual approach, put the disrupter charge alongside it and 'blow it apart'.

He listened thoughtfully as I told him that though I agreed with the manual approach I did not think much of his plans for disruption.

'As you've made up your mind to go for a walk, you can do that. But you will attach the hook and line to the device and we'll pull it out together. We'll decide what to do after you get back from your first trip ... but I want that bomb intact.'

At that time I did not know that this was to be the only live device that he dealt with on his four-month tour, not that it would have made any difference to my decision. I did make sure, however, that he got all the credit for the clearance action. That ensured that he earned his 'Felix' tie. Anyway, he was going to do the dangerous bit and make the first manual approach to a RC device.

We moved the EOD vehicle into the road leading to the field

* Prisoners who wore only a blanket wrapped around them to draw attention to their campaign for political prisoner status.

where the bomb had been planted and the staff sergant got into his bomb suit. After a few minutes he set off and disappeared through the gate into the field. On his reappearance shortly afterwards I broke the rules and, though wearing only an old sports jacket and cords, went forward to meet him, consoling myself with the well-worn argument that as I made the rules I could also break them.

We moved back abbut 50 yards from the gate and could feel the line taking the strain before it gave way. At least we had moved something. Now it was my turn to have a go. I took a careful look at the blank windows overlooking the field, cast a swift eye at the escort, and then made for the gate. In my view it was essential to get the thing over with quickly and I didn't waste time by climbing into the staff sergeant's bomb suit. In fact, our hook and line had worked and the device, a one-gallon water container, had been pulled away from the hedge, spilling some of the contents. A box commonly seen used by workmen to carry their lunch lay beside it. Only a few swift snips with my knife were needed to cut away a standard MacGregor radio receiver and encoder in a blue alloy casing, and the timing and power unit and the detonator. We had recovered complete a radio-controlled bomb!

It was hardly a pretty thing. The water container held 14 lbs of commercial explosive and 2 lbs of shipyard confetti (coach bolts mainly) with just for good measure, 30 1½-volt dry batteries added to give an extra shrapnel effect. The bomb was powerful enough to have killed or severely injured a four-man foot patrol.

I calculate that during my 14 months as CATO Northern Ireland only about one per cent of the bombs with which we had to deal were radio-controlled. The percentage has increased considerably since then with the decrease in bombing generally and with the terrorist element trying to develop its potential. It is comforting to know that with the evidence provided by EOD teams and forensic experts, the scientists are constantly working on improved counter-measures.

CHAPTER ELEVEN

An ATO's Last Words

We sat each surrounded by a cocoon of silence within the noise-filled cabin of the Gazelle and did not try to talk through the static which filled our headphones. Occasionally the shadow of the racing helicopter flickered over the desolate winter landscape as the sun broke through for pale moments but otherwise the countryside seemed deserted. Just what had gone wrong? What would we find when we arrived? Those were the questions but it was pointless to guess at any answers. In any case, we should be on the spot soon enough. As we neared the LZ it occurred to me that only a short time previously Sgt Walsh himself had been looking down on the same scene from a similar aircraft and wondering what he was going to be faced with when he landed.

It was a Sunday and, for me, had started normally. It was my custom at the weekend, if I had no specific task, to drive round the Province and visit each of the EOD teams, or at least as many as was practical under the circumstances. Coupe had picked me up at the Mess at nine o'clock and I gave him the itinerary of the day – Bessbrook, Armagh, Omagh, Londonderry and Magherafelt. I had decided to give Lurgan a miss and there was nothing going on in Belfast. Before we set out I checked that Coupe was armed, like myself, as we would be travelling through some problem areas. Having told the ops room where we were going we set off straight down the A1 to Newry where we turned off over the Newry Canal (reputed to be the oldest in Europe) and past the Derrybeg estate with its Republican tricolour futtering in defiance of the authorities.

The estate had spawned a lot of unpleasantness and evil and it was always pleasant to see the back of its smugly secretive semis and drive up the winding road to the corrugated iron gate of Bessbrook's grey mill where the sentry could be relied on at least to pass the time of day cheerfully. It was comforting too to park up alongside the EOD vehicles which stood in a businesslike row in the mill yard, pointing at the exit ready to go and obviously under the control of men who knew what they were doing.

The mill itself was a gloomy rambling place with iron staircases climbing here and there in an aimless way and a jumble of partitioned rooms where electric lights were needed permanently. Having checked into the ops room, which had been warned of my arrival, I asked if they would let me know when Mick, the weapons intelligence officer and an old friend, arrived; then I went off, mole-like, along the pungent trail that led to the burrow occupied by the EOD team. When the stink of CO-OP became overpowering one knew one had arrived. There was little ventilation so the smell of samples of explosive taken for forensic purposes was ever present.

The section occupied a room which had a common area lit by a solitary light suspended from the ceiling and around the sides were cubicles with blankets hung across the entrances to give some degree of privacy. In one corner of the general area was a TV set which the team rented privately, each man paying his share, and to one side was the electric kettle with a jumble of mugs and makings.

The section commander, RQMS L—, emerged sleepily from his lair and, after coffee had been produced, I sat on a battered settee alongside the signaller and the No 2 while we chatted about equipment and problems – and there were very few problems with EOD teams – and tasks that lay ahead. When Mick turned up a few minutes later he and the RQMS and I went off to the ops room to talk to Bryn Campbell, commanding the 1st Royal Highland Fusiliers, about the 20 or so suspect devices waiting to be cleared in South Armagh. Like the petrol tanker we had dealt with on Christmas Eve, only a fortnight earlier, or a railway bomb, clearances would be carried out only if the device interfered with patrolling or free movement of the public. There was nothing of this nature about and having finished our discussion I accepted an invitation from Mick to have lunch with him at 3 Brigade HQ. Things seemed so quiet that I decided not to go on to the other stations I had intended to visit and got L— to advise them. About 12.30 we swung out of the Mill following Mick's car as

140

he headed for Portadown. Life appeared quite civilised and I began to look forward to lunch and a further chat to Mick, an RAOC captain who had won the George Medal on a previous tour as an ATO in Northern Ireland. It was a complete surprise when, on the Market Hill road, he pulled over to the left and stopped. Coupe pulled up behind him and I got out and walked ahead to see what was wrong. Mick was listening to a crackling message on his car radio which was tuned into the 3 Brigade net.

'What's up, Mick?'

'It seems that one of your operators has been injured in an explosion, Colonel. I've no more details but the Brigade major is waiting for you at HQ. He'll be able to put you in the picture.'

We went on at high speed to Portadown and 3 Brigade's ops room which, instead of being cheerfully busy, was unusually quiet. H Jones, the Brigade major, greeted me with the news that the EOD operator at Omagh had been carrying out a clearance near Newtownbutler that morning when the device had exploded.

'We think your man may have been killed,' he said.

There were a few more minor details but that was all.

'OK H ... thank you for putting me in the picture. Have you got a helicopter I can use?'

'There's one standing by waiting for you. I assume you will want Mick to go with you?'

'Yes, please.'

I asked for the grid reference of the incident control point so that we could fly in direct instead of having to go via Omagh which would be a waste of time.

'And you might ask A— (commanding 3 Bde EOD section at Lurgan) to meet me at the ICP. Oh, and tell HQNI where I'm going.'

'Right, Colonel. We'll get on with this and we'll see you later.'

Mick and I left almost immediately for Newtownbutler but not before I had sent Coupe with the car back to Lisburn with a message to S—, commanding 321 EOD Unit, to stay put at HQ Northern Ireland to deal with any problems that might arise.

The area of the incident was easily identifiable from the air,

141

with the Ferret scout cars of the 9th/12th Lancers, then serving an 18-month tour, blocking the approaches to the vincinity and vehicles and men from the clearance party tucked away in tactical positions. The chopper went down smartly and a soldier from the covering party waved us towards a small group as we scuttled away crouching under the whirling blades. The Gazelle took off again immediately. A familiar figure was standing among them – George Vere-Laurie, then commanding the 9th/12th. He had been at Shrivenham as a degree student when I was on my long ammunition course and later we had been in the same division at the Staff College. There was no need to stand on ceremony.

'Hello, George. What's happened?'

There was little to tell.

'Sergeant Walsh was carrying out a clearance this morning. He was dealing with the device, apparently, when it exploded. No one has been forward since the explosion in case there is anything else down there. I came out here as soon as I heard there had been an accident, but his escort can tell you more about it.'

I said I would talk to the escort and then set about clearing the area.

As it happened, I knew the man quite well, a quiet and taciturn lance-corporal in the 9th/12th who was one of the permanent escorts to the ATO at Omagh, in this case Sergeant Walsh. He was obviously affected by the incident, but very much in control of himself. He was anxious to help and quite lucid.

'Sergeant Walsh and I came down together by chopper this morning, sir. He told the rest of the team to follow by road and meet us here. He wanted to recce the area first and save time.'

Walsh, it seemed, had taken his hook and line and demolition kit with him. When they got to the incident control point he decided to go ahead and see what the problem was instead of waiting for the rest of the team to arrive.

'But did he tell you what he was doing?'

'Not to start with, sir. He appeared round the corner of that building down there and told me that he had got the device out of the place and had put it behind that low wall you can see on the right. It was a milk churn.'

I looked down the road which ran straight towards the

Republic.

On the right was a house at right angles to the road and beyond, well back and invisible from where we were standing, was a grocer's shop butting onto the house. There were petrol pumps near the road, with a big puddle beyond them. From the air the buildings resembled an 'L'. The low brick wall ran for a short distance at the end of the house parallel to the road. Opposite was a thick hedge.

'What happened next?'

'Sergeant Walsh shouted that he could see what was inside . . .'

'And what was inside?'

'He said there was some cordite running through the explosive and he reckoned he could deal with it. I could see him putting his hands inside the churn.'

The escort had been covering Walsh from a gateway about 50 yards away.

'After he had been fishing about inside he shouted he was almost at the bottom . . .'

The lance-corporal paused and looked at me in a slightly baffled way.

'Then he simply said, "Oh, Christ, what've I got here . . . ?" and there was a bang.'

The young man in the combat suit kept his eyes on me and waited patiently as if for some explanation. He had seen it all and told it as he had seen it. Now it was up to me, and all I could say was, 'Thank you, corporal.'

I asked him if he felt fit enough to wait a little longer in case I wanted further words and he replied resolutely that he would be OK. He had seen a lot of death in Northern Ireland.

I went forward alone to inspect the scene of the explosion, taking the route used by Walsh. There was nothing to see by the side of the wall, not even a crater, I went into the store and checked all round. Once again a blank. Having satisfied myself that there was nothing else in the immediate area I went to the corner of the building and called for Mick to come and help. He looked at me quizzically and I shrugged.

'There's nothing else here, Mick,' I said. 'We'd better see if we can find Walsh and whatever it was that killed him.'

By this time A— (the EOD officer from Lurgan) a warrant officer and a forensic scientist from DIFS had turned up and we organised a sweep across the fields to find Walsh, the remains of the bomb and the pistol which Walsh had been carrying.

Only those who have taken part in such an operation can really imagine what it was like. Mick went off to borrow a broom from the distraught shopkeeper and began to sweep the water out of the puddle in case any bits of bomb had landed in it. The rest of us got hold of the bags used by DIFS to collect forensic evidence and began to scour the area. From time to time one of us would call for the broom to hook an object out of the hedge. What was left of the detonator of the bomb was recovered and the mis-shapen pistol. Finally Walsh's remains were found where the blast had hurled them, 150 yards away.

The search having been completed, there was little else to do. I instructed A—, a neat young officer who oozed confidence, to take the EOD team back to Omagh and to arrange for an ATO from Lurgan to join them immediately. By then it was bitterly cold and by the time I got a lift to the RUC Station where the 9th/12th had a troop, it had begun to snow. A white sheet was drawn over the quiet country road and the isolated houses where Walsh met his death.

(About two hours later, in the most appalling conditions, a helicopter put down at the rear of the RUC station which seemed to be surrounded by telegraph wires. I made the mistake of commenting on this remarkable piece of airmanship later and John Williams, the Army Air Corps colonel, promptly went in search of the pilot to charge him with flying below the safety limits. Fortunately for the intrepid flyer I had forgotten the details of the incident when asked later.)

S—, the OC of 321 EOD Unit, had arranged for a meal to be left in my room and when I got back to Lisburn about 8 pm he joined me there to tell me what had been done about the casualty procedure. I authorised every operator to phone home to tell their wives and families that they were not to worry when they heard the news on the radio or television. The only wife not to get that telephone call was Mrs Walsh. In her case the military next-of-kin procedure was carried out.

I have heard since that it had been broadcast once before that an Army bomb disposal expert had been killed in an incident, causing great anxiety to the relatives of people serving in Northern Ireland. If so, the precaution of getting operators to phone home was doubly necessary.

Later that night I phoned Mike Newcombe and Brigadier Lawrence–Archer, Director Land Services Ammunition, to ask for a replacement and, of course, to let them know what had happened.

What had happened? Let us deal first with Walsh, whom I had met personally only two or three times. Sergeant Walsh was a slim, cheerful young man of medium height, who had done previous service in the Province with the Royal Military Police. He had been impressed by the performance of the EOD operators and had transferred to the RAOC and attended at AT course. Subsequently he went to Northern Ireland as an operator.

He was a self-assured young man – confident and liked by his team. He was also impatient. Normally these are not traits which would kill an operator. Unfortunately a collection of circumstances were against him.

First, he had been back from rest and recreation leave only one full day before he went out on the clearance operation. So he had not had time to get himself back into the frame of mind necessary for the EOD operator's tasks. Don't forget, he was a confident young man.

Second, he had made some very successful clearances. Unknown to us at the time this included one when he had knocked the lid off a milk-churn bomb with a hammer.

Third, he did not go to bed early the night before the clearance.

Fourth, the inexplicable had intervened. As someone said later. 'What the hell was he doing putting his hands inside the milk-churn full of explosives?' A drawing of the boobytrapped milk-churn I had neutralised beside the tanker on the border on Christmas Eve was actually on the wall of his office in Omagh!

Furthermore he had been asked by the RUC Special Branch to visit them before he started the clearance. They had some information for him on the bomb ... which he did not call to receive.

Finally, the owner of the shop, who had found the bomb behind the counter, had advised him to blow it up *in situ*. It was good advice. But he did not take it.

He had not even worn his bomb suit (though it would

not have saved him). Under the circumstances of country clearances it was known that bomb suits were not always worn, but this was still a requirement under the rules. What makes Walsh's action even less explicable is that he was aware that WO P—, his diminutive predecessor at Omagh, had once cleared six milk churns from a customs caravan and that one of these had been booby-trapped in the same way as the one I had cleared near the petrol tanker on the border. But how can one explain the thinking of a highly-trained professional technician who flouts all the rules designed to protect himself and others when dealing with things that kill?

The only action that was correct, and even that was open to some question, was to drag the milk churn out of the shop on the end of a hook and line. It seems pretty certain that he did intend to deal with the device in the time-honoured way, placing primers on the base and blowing out the contents with a controlled explosion. In dragging the churn behind the little wall he may have had as his purpose the protection of the petrol pumps. It could be argued that this was a sensible thing to do. Had he used primers and retired a safe distance he would have achieved another success. Instead something made him change his mind with terrible consequences.

The churns which had been left by the car beside the tanker had been carefully planted to trap the unwary, which is why I circulated a description of it immediately. That alone should have made operators doubly suspicious. Nevertheless, despite his errors, which are easy to comment on when you have not had to do the job yourself, Sergeant Walsh was a very brave man. However, instead of a medal, Sergeant Walsh simply got a premature epitaph which should be burned into the minds of all EOD operators – his last words. Noted for posterity by a lance-corporal in the 9th/12th Lancers they are a warning to us all . . . 'Oh, Christ! What have I got here?' If ever you have to ask that question it is too late.

The Walsh tragedy had an unpleasant aftermath. The sergeant who took over immediately did a fine job in holding the Omagh team together and rebuilding their morale. (He and I did a clearance in a snowstorm shortly afterwards.) Then he was replaced by a staff sergeant from England who spent two days with us at Lisburn. He was given every assurance and told there were no bogey-men in Northern Ireland.

Provided that he stuck to the rules, all would be well. He was sent to Omagh to do a clearance operation which had been held back specially for him and after he had completed it successfully he was left to take command of the team. That same night he was on the phone to the OC of 321 EOD Unit.

The conversation must have been the most painful to which an officer in charge of a unit can be subjected. The man on the other end of the phone broke down and confessed that he was frightened – so frightened that he could not do the job. Coming from a senior rank that was a very serious admission. But it was serious also for the troops serving in the area and for the public, too, for that matter. For the second time in a matter of days the EOD team was leaderless and the area was exposed. The OC of 321 EOD Unit had no option but to order another NCO to take over (the sergeant who had done such a good interim job) and to send home the staff-sergeant who had cracked. In less sympathetic days, the staff-sergeant might have been dealt with severely – the kindest comment from operators in Northern Ireland at the time was that he should have been court-martialled for 'cowardice in the face of the enemy'. Someone suggested he should have been sent to a psychiatrist but we didn't have one in the Province at the time. In any case, when he saw one later he was found to be perfectly normal, and I am pleased to say that later he returned to Northern Ireland and completed a four-month tour satisfactorily. Perhaps the one thing one cannot fault is the man's decision to phone his OC – an act which, in itself, must have taken some courage. But having done so, the OC had no alternative but to relieve him. He could not possibly have left a man who had broken down to lead an EOD team in to a deadly area, or any other area for that matter – for their sakes as well as the man's.

Unhappily the Director Land Services Ammunition was highly critical of the way in which the incident had been handled. Why had the man been sent to Omagh as a replacement for Walsh? Why hadn't he been sent to Londonderry or Belfast where he would have had other ATOs around him? One of these should have been sent to Omagh instead. Undoubtedly there is some merit in his argument. But ATOs cannot be kept safe from things that go bang in the night – they can only be protected by their training and experience and character. Or they should be. Otherwise they should not be sent to do the taxing job they face. Moreover, in the opinion of the commander on the spot, S—, there had been nothing apparent to indicate that the particular senior NCO was unfit to lead a team in a country location. Indeed by transferring an ATO from an urban location, such as Belfast, one might be subjecting him to unfair strain because of the change in environment, the difference between country and town operations being marked, as I have discussed in a previous chapter.

I fully endorsed S—'s action then (as I do now), but the affair soured relations between myself and the Director for the rest of my tour.

The people who felt the death of Walsh most were S—, as his OC, and Sergeant A—, in the Belfast team. Walsh and Sergeant A— had served together in the same section of 1 Ammunition Inspection and Disposal Unit in England and had gone out to Northern Ireland together. They were pals and only one of them would be going back.

The EOD operators, the ATOs, were only too aware of the danger of their job. In the early days of the campaign, before selection procedures and training and selection were tightened up, the casualty rate was as high as 20 per cent – i.e. one in five was likely to be killed or injured. By 1976 this ratio had fallen to five per cent. As there were 20 operators in Northern Ireland at any one time, including myself, that meant that one of us would probably be killed, still a high risk.

Operators had their own ideas about survival, and a few asked not to go home for R and R during their four-month tour. I fully supported such requests. Personally, I used to find it extremely difficult getting back into the right frame of mind when returning from the many trips I had to make back to England. For an individual operator returning from leave the situation was the same. You arrived in England still keyed up and alert but by the time you were on the way back the sharpness had gone.

In my opinion, the time when an ATO was most at risk was just before he went on R and R, just after leave, and just before he was due to leave the Province at the end of a tour. This meant that for three of the sixteen weeks he spent in the Province the odds against him increased. Because of this, the OC of 321 Unit and I used to try, although it was sometimes impractical, to accompany operators on clearances during those periods. As far as newcomers to the theatre were concerned, the terrorists did not oblige us by informing us in advance of their plans so that we could break in people gently. Besides, there was no such thing as a 'safe job'.

Moreover, it would have been unfair and wrong to expose the best operators to the greatest dangers all the time. Their chances of being killed would have increased enormously. Finally, I had the distinct impression that the breakdown of the staff-sergeant who was sent home was laid at our door because the training and selection of personnel at home, of which DLSA was the arbiter, must not be faulted. The operational environment, however, was quite different from that of the lecture room and there was no way in which those in control in Northern Ireland could discover a weakness in a particular individual – officer, warrant officer or sergeant – until he had been exposed to reality.

What needs to be borne in mind was the high quality and cool courage of so many of the operators and what they achieved in four months, whether it be Staff Sergeant B— with his clutch of mortar bombs at Crossmaglen, or Warrant Officer S— (a man so small that he seemed to disappear when he got into his bomb suit) neutralising six drums of booby-trapped explosive.

CHAPTER TWELVE

A Man Called Mac

In the files relating to the Northern Ireland campaign there will be, by the time it is over, a mass of official foolscap forms carrying a bureaucratic tag relating them to the work of the EOD sections. They are the IED (Improvised Explosive Device) Incident Report Sheets ... and it is to be hoped that no-one ever loses the art of reading them.

The problem that would face any future historian is that they are written in specialist jargon, a sort of EOD shorthand. Scanned by an expert they can reveal what time a man was tasked, when he went out on call, the time he completed his operation, the location, target and all details relevant to the type of bomb, the principal target and so on. It is also possible to follow the operator's actions – and that is what makes these otherwise dull-looking ordnance branch forms so fascinating. They reveal the thoughts in the mind of the individual on each occasion, as well as what he did as he risked his neck.

Because of their official nature, the space reserved for the comments of operators contains usually only a terse phrase or two. To put down what went through the heads of some of us after a particularly hairy operation would not have made particularly good reading. It would not often have been even decent. No budding authors will emerge from these reports. Had each operation been described in full no one would have escaped from their desk, or scrubbed table or whatever. No-one had the time to write down – 'Thought about it for a bit ... said a short prayer ... called in search team ... said another prayer when they had confirmed the report....' No, the description of the action was short and to the point and the only time one was likely to come across a literary touch was in the final comment column. (A baffled warrant officer who neutralised a 2½ lb bomb on top of a cistern in the ladies' lavatory of an hotel in Newry declared that the reason for the position of the device was not apparent – 'unless the bomber has a loo fetish'.)

One man of very few words was WO1 McKernan, who was serving in Belfast. What is remarkable is the number of times his signature appears at the foot of an incident report, especially as Mr McKernan should not have been in Belfast at all according to the book. He was beyond the normal age for an operator and could easily have avoided a posting to Northern Ireland. However, he was a rather old-fashioned soldier and insisted on doing his duty. In his eyes it would have been a stigma to have completed his service without having carried out the sort of operational tour that many of his subordinates and friends had undergone.

Though I do not agree it would have been a stigma – clearly not everyone who is qualified can do a tour – one has to admire the spirit of the man. I also learned to admire his courage and ability, but how these could be discerned by any historian researching the files in years to come, it is difficult to see. Let us look at a few days in the life of this veteran as revealed by expert reading of his monosyllabic reports.

The date: 4 December 1976. The place: Winsor Park Football Ground, Belfast. The operation: to neutralise a 60 lb device left on the back seat of a car which had been hijacked in the Andersonstown area the previous day. A routine foot patrol of the 2nd Regiment, Royal Military Police, had spotted it, wisely called an ATO, and left it to him. The bomb had been left on the back seat of a Datsun and contained home-made explosive with a clockwork timing and power unit. Five pounds of ballbearings had been added to give a deadly shrapnel effect if it exploded. The operation took the ATO two and a half hours, and he described it completely in little more than twenty-five words.

The date: 5 December 1976. The place: Orient Gardens, Belfast. A Ford Cortina had been found, apparently having crashed while being used by terrorists, possibly as a get-away car or a vehicle on the way to a hit – for some reason the Cortina, preferably 'dirty white', was a favourite of gunmen. The driver's door hung open and there were ominous red smears on the upholstery. In the back of the car was an anorak partially covering a .22 rifle.

The troops who found the car – gunners of the 32nd (Light) Regiment, Royal Artillery* – took no risks. Nor did Mr McKernan.

Only after Wheelbarrow had been used as a camera and a weapon, and the boot was blown open, revealing nothing, did he make a manual approach. Judicious application of hook and line on the contents of the car neutralised a booby trap device under the anorak. Anyone reaching into the car to handle the .22 rifle would have set off 3 lbs of commercial explosive packed neatly in a plastic bag. The smears on the driver's seat turned out to be tomato ketchup. Time for the job: one and a half hours. Comment: none.

The date: 5 December 1976 again. The place: Ramon Gardens, Belfast. WO1 McKernan was called out at nine in the evening after a routine patrol of the 1st Royal Welch Fusiliers found a suspicious car. Yet again the barest minimum to describe a four-hour task. This time a command wire was discovered and traced to a firing point in a garden almost 100 yards away. After using Wheelbarrow to look all round the vehicle this indefatigable old soldier forced a remote entry and then walked up in his bomb suit to attach hooks which removed, eventually, a milk can which contained 12 lbs of powerful explosive *plus* 30 lbs of shipyard confetti, nuts and bolts which would have sprayed the area if the device had been allowed to explode.

The date: 13 December 1976. The place: Wilson Street, Belfast. On this day, which is memorable for the number of devices and hoaxes planted province-wide, Mr McKernan carried out six tasks. The first, which took him five hours, began when he was called to a brush factory in Wilson Street, where terrorists had broken in, shooting down and killing one of the staff, and done 'something' with devices before escaping. This time the problem was complicated by the inability of Wheelbarrow to get at the most obvious device – a carrier bag left in an office downstairs in the factory. Once again it had to be an approach in a bomb suit before the device, which turned out to contain 20 lbs of explosive, was disrupted and neutralised.

As two other devices had been reported, in the same building, the whole painstaking business had to be gone through

* Now 32nd Guided Weapons Regiment, Royal Artillery.

again before they could be declared as hoaxes and the premises declared free from threat.

The date: still 13 December 1976. The place: Merville Gardens, Belfast. Towards midnight, WO1 McKernan was once more in his bomb suit, now smelling even more strongly of explosives and sweat and not getting any lighter. By torchlight he neutralised a 10 lb bomb left in a carrier bag in an office block doorway. On his report, he reduced the whole operation to seven words. There was no comment.

Three weeks later the same warrant officer once again responded to a number of calls on the same day (4 January). In a store in Landseer Street he neutralised 24 lbs of CO-OP with a gallon of petrol attached, he disrupted a 14 lb bomb in a funeral parlour in Bedford Street, and disposed of a 7 lb bomb in a car found in Welsh Street. In the first two cases, he was unable to use the Wheelbarrow and had to approach the devices on foot and hook them out on the end of a line.

It was a remarkable performance and he richly deserved the award later bestowed upon him. Such men as he could take pride as they filled in another section of that austere report form ... the line which required the operator to describe his 'Rank or Profession'. Two simple cyphers were all that were needed – in his case 'WO1 ATO'.

Mr McKernan's fearlessness, by the way, was not confined to bombs. He was a Scot of few words and firm opinions. Known to everyone in 39 Brigade, including the brigadier, simply as 'Mac' he did not hesitate to give his superiors the benefit of his opinion if he thought it necessary. I know because I was in his 'patch' when he was operating in Belfast on 4 January. I was moving around from team to team that day and watched him in action two or three times. By some freak, however, when he was called out to the funeral parlour, I got there before him and actually found a witness who had seen the bomb planted. To save time – and there was so much going on that day we weren't leaving anything to soak – I took what evidence I could and then, dressed in my tweed jacket and flannels, went in the building to have a look around. When Mac arrived I was waiting by the door.

'Beat you to it this time, Mac,' I said as he got out of the vehicle.

He grunted.

'And I can tell you where the bomb is.'

His eyes narrowed.

'Come on and I'll show you.'

I began to walk towards the door of (I think) the carpenter's shop.

'Oh, no you won't, sir,' came the reply. 'How do you know where the device is?'

There was no way out. It was quite clear to him that there was only one way I could know. I must have been inside – which I had. I had discovered the bomb at the back of the workshop in the middle of a jumble of planks and wreaths. I tried to be casual as I explained, but it didn't wash.

The old warrior looked at me with fierce eyes and choked back his anger.

'You must be bloody daft . . . sir.'

His mind was made up and there was no way anyone would change it.

'I'll deal with this device,' he said. 'Properly.'

With that he returned to the vehicle and put on his bomb suit and dealt with the thing.

Mac, of course, was quite right. If we had gone in together and the device had exploded it could have knocked out two operators. Mac's commonsense gave me a jolt. Up to that time not one device had exploded of its own accord during operations in which I was involved and I was beginning to stray down the path of fatuous belief that nothing would happen ever. It was better to be put in one's place by a warrant officer with a short temper than a device with a short fuse.

Some of the incident report forms tell a story of real conundrums encountered. Sergeant S—, working in Belfast, was set a task by a particularly perverse terrorist in September 1976. This artist placed a device inside an armchair in a shop in Ligoniel Road and then pushed the chair into a lift which he managed to jam between two floors. It took the sergeant two hours to lower a disruption device onto the chair from the top of the lift shaft. The armchair promptly caught fire. Later when he was probing the ashes the detonator functioned but by then the device was no longer viable. It was just as well – it consisted of more than 5 lbs of commercial explosive.

A few weeks later the same NCO was tasked to neutralise a bomb believed to be in a Land Rover which had been hijacked by a gunman and driven to a garage on the Lisburn road, where the terrorist carefully locked the doors and went off with the keys.

I was at the scene with Sergeant S— and we suspected the bomb was in the tool box under the passenger seat but we didn't know how to get at it. All around us were gleaming Audis and Mercedes, and we didn't have much room to manoeuvre Wheelbarrow. The sergeant sent for some skeleton keys and while we were waiting for them to arrive he tried to open the door with plastic explosive. The only result was to jam it still more firmly on the passenger side. This meant that when the skeleton keys did arrive he had to get in the driver's side and lean over inside the vehicle to lift up the seat. All the activity that had been going on around the bomb – two Wheelbarrows had broken down on the job and we had to get a third – was likely to have made the device more sensitive so when S— eventually peered cautiously inside he was fairly keyed up. It was rare for this ATO to add anything to the technical details in his after-action report but for once he broke into print – 'Device in tool bin – BEAT RAPID RETREAT,' he wrote.

As he was eyeball to eyeball with 65 lbs of explosive and five gallons of petrol, the more rapid the retreat the better, in my opinion. The device was finally cleared after five hours with only minor damage to the other vehicles in the garage. We were all pretty tired by the end of this operation.

Down at Warren Point a watchful RUC constable looked under his car and WO2 L—, the local ATO, was called in to deal with a pretty little problem. A most peculiar object was attached to the suspension of the offside front wheel. Use of Wheelbarrow and shotgun techniques eventually enabled him to clear a bomb containing six large sticks of explosive in a box which had been attached to the car by a number of powerful horseshoe magnets. The only damage to the car was a punctured tyre and a hole in the wing.

St George's Day 1977 saw another warrant officer ATO dealing with a very complicated booby trap. An explosion had been reported near a border crossing at the beginning of April. When the Security Forces arrived they observed a large crater by the side of a road and not far from it lay a cable reel and batteries. On the other side of the road lay, so first reports said, a human arm. An 'own goal' perhaps? In the light of the booby-trap campaign which had been waged in South Derry earlier that year a planned operation was initiated and it was not until the incident area had been well observed for some days that it took place. A diligent search team finally located two trip wires. Anyone making a sweep to find other human remains would have triggered off two improvised explosive devices, both of which were dealt with by WO2 A——. One of them contained 45 lbs of home-made explosive, the other a total of more than 70 lbs of explosive. The 'arm' turned out to be the sleeve of a combat jacket and a light coloured glove, both stuffed with soil.

A fortnight later A—— was filling in another report describing how he cleared an 11 lb bomb hidden beneath a catwalk on top of a fuel tank containing 12,000 gallons.

The previous August another operator, Sergeant C——, had found himself poised within a few feet of a tank containing 2,000 gallons of petrol in the premises of a car hire firm. Originally he had been told by eyewitnesses to the planting of the bomb that it had been placed behind the door of the ladies' lavatory but by persistent questioning he established that there might be 'something' under the stairs. Getting into the ladies' was a problem because of the narrowness of the door and eventually he had to make a manual approach. It was then that he saw the actual device, a 15 lb bomb with a gallon of petrol attached to it, under the stairs. This meant a very delicate clearance in the knowledge that any slip would turn the place into an inferno. But he succeeded.

Two days later half a dozen gunmen entered the Ardoyne bus depot and taped 11 devices against the sides of vehicles. Three of the bombs exploded, doing some damage to two buses. But Sergeant C—— dealt with the eight remaining devices, some with Wheelbarrow and shotgun. The others, in places that could not be reached mechanically, he dealt with by hand. The meticulous Scots major then commanding 321 EOD complimented C—— on a job well done but noticed a flaw in his technique as revealed by the report and gave him a rocket for the lapse. There was no question that everyone was in a position to learn – and close scrutiny of the reports by an expert like Jock could only help.

Some idea of the difficulties facing EOD teams in Belfast at

156

this time might be gained from another report filed by Sergeant C—.

On 1 June 1976 gunmen pushed their way into business and warehouse premises in Ormeau Avenue and planted two bombs. The sergeant's team arrived just as one went off on the first floor. After the dust had cleared the operator began his search for the second – on the ground floor – but the way to it was blocked by debris. The side door wouldn't budge to the tugging of a hook and line and the steel slatted main doors were jammed, having been twisted by the blast. Eventually C— opened the latter by improvising his own hook and line – a couple of hefty spanners forming a grapnel on the end of a tow rope attached to Wheelbarrow. The result was a partial success only. After the doors creaked open, Wheelbarrow and its camera were able to get in so far but rubbish and fallen goods prevented further entry. The sergeant went back to the side entrance and began clearing rubble. When this had been done he climbed in again in his bomb suit and used disruption devices. These simply shifted the bomb to an even more difficult position. Once again this persistent operator reviewed the situation and came back to the attack via the front door, having to move a massive bale of goods to get to within six inches of the device.

The obstinate struggle took some six hours on a sultry June afternoon with the operator sweltering in his bomb suit, sometimes staring at his target through the closed circuit TV lens, often puzzling over the next line to his quarry, a cardboard box about a foot square. At the end of the day it was the operator who won. The bomb, containing 15 lbs of explosive attached to a gallon tin of petrol, surrendered its secrets for the attention of the forensic evidence experts.

Many stern and dangerous struggles like those I have mentioned have been fought out by the men of 321 EOD during the current troubles, lonely battles, sometimes in the dark, generally with the smell of explosives in their nostrils, their hair sweat-soaked inside their heavy helmets and their darting eyes always looking for the unexpected, the trap, the tell-tale sign of the thing that kills. It may be as well therefore to review the way in which the NCOs and warrant officers were selected for duty at the time of which I am writing. I cannot pretend I was entirely happy about it.

157

At that time, if you were a sergeant ammunition technician in the Royal Army Ordnance Corps you had to go to Northern Ireland if posted and passed through the selection system. If you didn't you lost your trade and had to remuster as an ammunition storeman or some such lesser grade. You lost caste, prestige and pay.

I do not believe this was necessary. A man may be good at his trade but when the extra element of danger comes into the application of it, surely it does not make him intrinsically less skilful, though it may make him less effective. A man may have an outstanding sense of hearing and direction, but not everyone with those qualities is prepared to put his fingers in his ears and cross the M1 blindfold.

The soldier on internal security duties in Northern Ireland stands as much chance of being killed as another member of the SF carrying out similar duties. Generally speaking, when a four-man foot patrol sets out each member takes the same risk of becoming a casualty, because the terrorist sniper is not interested in a specific individual but a group target.

As far as the survival of the EOD operator is concerned, however, there is a direct relationship between cause and effect. The random element is vastly reduced. A small amount of luck may come into it – an EOD operator could be run down in his own garage – but to a great extent his actions will decide his fate. It seems to follow, therefore, that the EOD operator has to have an additional psychological requirement above that required of an expert technician. In civilian life, for example, a man may be required to control a large workforce or operate a complicated production line which has considerable relevance to the profits and smooth working of an important industrial undertaking. This responsibility he may accept readily and fulfil competently. But it does not mean that he would perform the same job in the same way if required to take a high personal risk, either financial or to life and limb. Indeed the ability to face up to high personal risk is not, in itself, the major criterion for the selection of senior management.

It is the ability to stand stress, while retaining a high degree of effectiveness, that is the prerequisite of people required to make major decisions.

In the case of the successful EOD operator, he must be able not only to stand stress but to carry out his operations without direct support undeterred by the fact that his nearest companion is likely to be 100 yards away from the neutralisation – the point of maximum danger during a manual approach.

A study of the qualities required of an operator has listed them as: good judgment; foresight; the ability to learn quickly from previous actions; a logical mind; the ability to be decisive and, at the same time, flexible; self-confidence allied with prudence; the ability to communicate; technical ability; a personality which commands respect from upper and lower ranks.

These qualities, which are those required of good middle and senior managers in industry and commerce, emphasise the ability to command but they are no guarantee that a person so endowed would be successful in an EOD situation. Two further requirements are necessary. A man needs to be self-contained but not introverted. In being decisive he needs also to be aware that his life and that of others may be dependent on his decisions.

A story told about the late Field-Marshal Lord Alanbrooke throws some light on the subject. Long after the Second World War when he was visiting the Staff College he told students how in 1937 he sat in the same hall with a lot of distinguished officers, all with high academic qualifications and reputations as intellectuals. I paraphrase:

'Then the war came and almost to a man my colleagues went out of the Army and we sent for the beer drinkers, men like . . .' and he named a top-ranking officer with an outstanding combat reputation.

In other words, technical qualifications and theory are not enough at the 'sharp end'. Character is what is required, and a deficiency in expertise may be more than compensated for by personality factors. If an EOD operator does not walk towards a suspect device he will not be killed. Commonsense dictates that he stays put. Therefore there has to be something extra in the operator's make-up that will make him advance and carry out his technical functions despite the hazards of which his brain has warned him.

In addition, the EOD operator must be able to 'live with himself' after he has completed some particular action. He needs to have privacy of a sort – hence the blanketed cubicles at Bessbrook Mill – and yet still be able to carry out another hazardous task when called out, perhaps within a matter of days or even hours.

Do such sterling characters exist in enough quantities in the Royal Army Ordnance Corps? The answer to this must be yes, because of the exceptional success rate in Northern Ireland, despite, at times, the very high death rate that occurred.

This could lead one to suppose that even operators with marginal or doubtful capabilities and personal qualities had proved up to the job after, perhaps, an initial settling-in period. It was with this supposition that I disagreed with the responsible authorities in England.

In the Northern Ireland context, certainly in the first seven years, it was not possible to lessen the risk to the doubtful operator. No matter what precautions were taken to put him in a low risk area, there was always the likelihood that a complex incident would arise – and then because of his failings he would probably be unable to realise that it was complex, let alone deal with it satisfactorily.

To get a true picture of the value of doubtful or marginal operators it would be necessary to relate their performance to the level of activity in their area; to consider changes in techniques and terrorist tactics; to relate their operations to changes in equipment; to be sure that the method they used to neutralise a device was correctly reported and that the correct procedure had been followed. The question of whether certain devices in the area had been dealt with by a senior officer because of doubts about the operator was also relevant.

Such an evaluation, though desirable in many ways, had become impractical by the time I was serving in the Province. Had a computer been employed from the start, had the reporting procedure been cut and dried, had the operators found time to do all the paper work involved, it might have been possible – but there was a war on and they were in the forefront. The only figures one could rely on were the numbers of occasions when operators failed and were either killed or removed from their duties. On that basis alone I argued that, at least, there must be doubts about the ability of the marginal or suspect operator to complete a successful tour of duty, and that it was not justifiable to arrange EOD postings so that a good operator faced a higher risk of death than a poor one.

I also held that a good ammunition technician who failed to pass the selection procedure (provided it was not due to gross

160

technical or disciplinary weakness) should not be penalised by losing his trade.

The fact that an NCO or warrant officer ammunition technician might not therefore have to complete an EOD tour should be viewed, in my opinion, in the same light that not all commissioned ammunition technical officers will complete EOD tours. That fact did not make them lesser officers in the eyes of those who had completed tours – why should it be a stigma in the case of NCOs and WOs?

These views did not find favour with the hierarchy at the Directorate of Land Service Ammunition.

Of the leadership of the captains who served in Northern Ireland as ATOs I can say only that I have the highest regard for these young men. During their tours of duty they had to command their sections, maintain morale, assume responsibility for all the equipment at their disposal, advise the brigade commander, ensure high standards were kept by the teams, lead their own team, and deal with bombs. In effect this meant that they worked flat out for four months. At times the strain on them, particularly in Belfast, was tremendous, and not every individual achieved the high standards required.

In retrospect, I can see that from time to time one came to accept their outstanding dedication and application as normal. When I heard that Captain B—, then commanding the Belfast section in Albert Street Mill, had gone into the Musgrave Park Hospital with 'ringing in his ears' my first thought was how to get him out. Belfast was pretty lively at the time and I needed experienced characters like him. I suppose I arrived at the Musgrave about 15 minutes after the patient, passing through the gate in the high fence erected against grenades and checking in my pistol at reception. The Musgrave was a little world in itself, with the windows protected against bomb blast and everyone working in the knowledge that occasionally snipers opened up on the Officers' Mess across the M1. Inside however, all was peace and quiet and it no doubt seemed like paradise to Captain B—. His sojourn lasted only one night, however, for after a hot bath and a good night's sleep he reckoned his ears were all right again and out he came.

In my view, as long as this country has young officers of that calibre serving the Queen we shall not go far wrong.

It was because of my admiration for the captain ATOs that

when Petrofina presented a silver cigarette box to the Corps, following the tanker incident at Dunmurry, I made sure the inscription began, 'In appreciation and admiration of the work of the Ammunition Technical Officers of the Royal Army Ordnance Corps in Northern Ireland. . . .'

CHAPTER THIRTEEN

The Gasworks Gang Go West

From time to time, I am asked, 'What exactly is an explosion?' The technical explanation, as far as I remember, is that it is the manifestation that follows the rapid breakdown of a chemical by a shock wave passing through it. It is the velocity of that shock wave that determines the difference between an explosion and a detonation, and that is called the power of an explosion. High explosives are those contained in detonators and include such substances as TNT, RDX, and plastic explosives. All these detonate. Low explosives such as ANFO, explode. It is the power of the explosion that determines the effect. One could reckon on CO-OP being about half the power of TNT and ANFO about one hundredth. The effect on humans of either high or low explosives is not so mathematically predictable, except that the scale of destruction varies enormously.

When a chemical breaks down as a shock wave passes through, it adds to the shock wave, building it up and producing a pressure front. It is this phenomenon, called over-pressure, that assaults the body. An over-pressure of five to eight pounds will perforate the eardrums. Ten to fifteen pounds causes the lungs to collapse. The more intense the energy release, the more devastating the effect. Pressure waves enter the mouth, nostrils or any orifice and disrupt the body from the inside; limbs are torn from the trunk; flesh may be stripped from the bone. All this happens in a fraction of a second. The speed at which a shock wave moves through an explosive chemical may be measured in thousands of metres per second.

Ways in which explosives are assessed include the Velocity of Detonation, Power and Figure of Insensitivity. A 'good' explosive has a very high V of D, high Power and a high F of I.

Detonators, which are unstable and used to set off other chemicals, are unstable and have a high V of D but low F of I. In plain terms all explosives are capable of inflicting terrible injuries on the human body, even in small quantities, and most of them are dangerous enough to command the greatest respect. Nor will age weary them. In the cases of some chemicals, the older they are the more sensitive they become. In July 1955 Belgian farmworkers left the field in which they were working to take shelter from heavy rain. Not long afterwards the field vanished, blown sky high by a mine which had been established at the end of an underground tunnel under the German lines near Plugstreet Wood, east of Ypres, in preparation for the battle of Messines in 1917. The mixture – the Messines mines averaged about 40,000 lbs of Ammonal, Blastine or gun cotton – had lain under the supposedly saturated Flanders mud for nearly 40 years and still had enough power to blast a crater big enough to swallow a village complete with church. No-one knows what caused it to explode though local legend says that lightning triggered off the mine which, until then, had been declared 'lost'. It is a sobering thing to know that another large mine, burrowed under the German lines at the same time, is still missing somewhere in the same region.

The sensitive nature of old explosives may be underlined by the casualties suffered by the special Belgian Army squads which operate still in what used to be the Ypres Salient and the coastal areas. In the region of 50 men have been killed since 1918, and sometimes as much as 300 tons of rusting shells and grenades are retrieved and blown up within a year. All explosive needs to be treated with respect and home-made explosive and improvised devices in particular. A grim example of what can happen to the amateur was presented to Northern Ireland in the middle of October 1976.

It was a Sunday and a number of officers had left Lisburn to attend a concert at the Ulster Hall. I mention this because afterwards one of them told me that though the lights were low and curtains drawn 'the whole place was suddenly lit as though someone had turned up a big old-fashioned gas mantle. Then we heard the bang'. A gas mantle it was indeed, but not the sort you could adjust.

From intelligence reports gathered later it appears that

three senior members of the Belfast brigade of the IRA had decided to have a go at the Security Forces base in the Ormeau Road gasworks. The main building itself is a splendid Victorian edifice: Sir John Betjeman, a connoisseur in such matters, once sat in a chair in the street opposite just to be able to gaze upon this jewel. But the whole area is a sprawling industrial site and to protect it thoroughly would require unlimited manpower, something the Army does not possess. Instead the place was being guarded by a detachment and patrols operating from a base within the works perimeter.

The field craft of the raiders appears to have been reasonable but their ordnance knowledge suspect. Having gained access to the gasworks they took cover amid some discarded old stoves in order to arm the bombs, of which there were probably two or three. Although we will never know for certain, it seems that one of the bombs must have short-circuited via one of the old metal stoves and what is euphemistically known as a sympathetic explosion took place. The whole lot went off with such force that one of the gasholders caught fire, causing a spectacular blaze.

I was required to inspect the scene and by the light of day saw at first hand the brutal, terrible effect of high explosive on the human frame. One bomber had been blown against a concrete post, one of many supporting a chain link fence surrounding the gasholder which caught fire. The force of the impact made by this body had snapped off the post at ground level. The legs were still attached to the body but it was without arms or head.

Some yards away there was a leg, complete to the thigh, then an arm, then more leg, then other unspeakable fragments. The welter of small pieces of flesh and clothing in one particular spot made it abundantly clear that one bomber at least had been 'strained' through the chain link fence – just as if he had been put through a mincing machine. For some time it was difficult to know how many had died in the blast but eventually the discovery of a backbone on top of one of the gasholders made it likely that three men in all had died.

Their intention had been to hang their bombs on the mesh of the windows of the post manned by our troops in the works, and it was while they were engaged in setting up the device that the short circuit occurred.

The dead in this case were later identified as leading figures of the Belfast brigade, one of them a quartermaster, another a training officer. This in itself was unusual, as it was normal for 'volunteers' to be used for the dirty work. Normally the men who selected the target, the men who decided when it should be attacked and how, were not the type to expose themselves to being torn limb from limb. They counted themselves too important to risk their own necks and they selected carriers from the credulous, the naive, the reckless, and the slow-witted to deliver their lethal cargoes. They sent off their minions callously, knowing full well that any number of accidents might cause their cargo to explode – a heavy bump on Belfast's uneven roads has been known to set off a bomb being carried in a car and it is not long since a device in a train exploded prematurely because of its faulty construction, once again taking a toll of the innocent. I cannot imagine what stories the bomb-makers tell the carriers to motivate them – no doubt colourful lies about exploits they are supposed to have carried out themselves – but they do not hesitate to use anyone for their own ends, regardless of sex or age.

How does one feel, then, when one sees the results of an own goal of the size of the Ormeau Road gasworks bomb? Apart from a revulsion at the visible horror and the pointlessness of it all, one cannot escape a grim sense that rough justice has been done.

The bomb is a weapon well-suited to a terrorist campaign – the results of an explosion are visible on the streets for a much longer period than, for example, a straightforward assassination. The numbers indirectly affected by attacks are large. Attacks may lead to works being shut, traffic being diverted because streets are closed, and families being forced to leave their homes because of the possibility of further bombs being found in a district. Best of all for the terrorists and their minions, in their own minds they can transfer responsibility to the ticking, power-packed, mindless monster they have constructed. If the innocent suffer, they have the audacity to call it bad luck or claim that they died for 'the cause'. According to one definition a psychopath is someone who is abnormally emotionally unstable, often with anti-social tendencies, but who has no specific mental disease. In the case of the indiscriminate bomber one must consider whether or not he or she is so consumed with hatred that it is an incurable affliction. Who but a homicidal maniac would have planted a bomb under the steps of the garrison social club where the wives and children of the battalion in Londonderry were at a party on New Year's Eve 1976?

In a free society, such as we enjoy at the moment, there are problems in trying to control the ingredients which can be used to make explosives. The 'Anarchist's Cookbook' was on sale openly until recently, specifying the type of bomb and quantity of material required to produce a specific effect. (One wonders where the terrorists responsible for the Bologna outrage in August 1980, when more than 80 people died and nearly 200 were injured, obtained their materials and bomb design.) If it is difficult to control the supply of ingredients used for producing home-made explosives, governments can, however, go a long way to prevent commercially manufactured explosives and detonators getting into the wrong hands. Both the British and Irish governments have done much to impose effective controls. After it had been proved emphatically that the principle source of supply of detonators to the IRA was from the Republic all detonators produced for use in the South were crimp-marked. All explosives manufactured by Irish Industrial Explosives Ltd are serial-numbered and, in addition, there are covert methods of marking. Similar precautions are taken in Great Britain.

The need for control, however, is international rather than national, simply because the terrorist use of the bomb is international. It is tempting to suggest that it follows that we should have an international bomb squad, but I don't think so. Much of the equipment that we use is classified – and we do not tell our secrets to the world at large. In doing so we could be handing information straight to the terrorist who threatens us. What Britain does do is contribute to a regular cross-flow of information between Canada, Australia and the United States, via bomb data centres. Unfortunately this system tends to be time-consuming and, during my service career, unless one was prepared to sift through interminable reports, it was all too easy to miss a new device.

While I was in Northern Ireland all the reports sent to Great Britain were held manually and though it had been suggested that, with relevant details from other countries, the information should be computerised, this had not then been done. This was a great pity, for if we identified what we thought was a new device in Northern Ireland it ought to have been possible to ask for a print-out of any similar devices used elsewhere in the world. It is to be hoped that this situation has been altered or is in the process of being rectified for, as current events show, terrorists of all colours, creeds and political beliefs are constantly improving their own techniques. Near Belfast itself the Maze Prison is unquestionably an advanced teaching establishment for the IRA bombers and at least two boobytrap switches have been produced by the prisoners there. The problem of breaking up this terrorist equivalent of the Staff College is a difficult one. How does one separate the bombers in the Maze? When they are not designing they will be instructing. The answer may be to split them up and send them to ordinary prisons elsewhere. But I am glad that someone other than myself has to find the answer to this problem.

One other aspect of the Northern Ireland campaign which did cause me concern when I arrived there, and which continued during my tour, was the imprecise structure of the intelligence-gathering organisation as it related to EOD. The need to establish a proper Weapons Intelligence Unit in the Province, tasked to identify future intentions of the IRA and trends in bombing operations, had been realised in the period 1973–4.

As the implementation of this requirement was influenced by the Army's professional nature, its promotion and posting system, and its tribalism, a few words of background may be helpful. Weapons Intelligence covers all the technical intelligence gathered rather than information on individuals, enemy formations and tactics. In olden days, of course, there was not the same need for specialists, smooth bore muskets and limited-range cannon being the basic weapons for 150 years. Today the need for technically qualified officers may be gauged from the plethora of weapons with which the private soldier in the infantry has to become familiar – the self-loading rifle, light machine-gun, general purpose machine-gun, 9mm pistol, sub machine-gun, recoilless rifle, wire-guided anti-tank missile, mortar with smoke, HE and illuminating rounds and sundry grenades. He must also learn to fight from an armoured personnel carrier, wear what he describes as a 'Noddy suit' to counter biological and chemical warfare, use a radio and be prepared to drive an assortment of vehicles. Private soldiers are specialists, so the need for officers trained to a much higher degree is essential.

Up to and including my tour in Northern Ireland there had been six CATOs of whom three, including myself, were Weapons Staff Officers having passed the technical staff course or qualified at the Staff College, or both. It was during the command of my qualified predecessors that the need for the intelligence unit was proposed and accepted.

The proposals were uncomplicated. A Weapons Intelligence Officer was to be located in each of the three brigades (3, 8 and 39). These officers, captains, would report direct to a qualified major at Headquarters Northern Ireland who would be subordinate to CATO. Thus CATO would be responsible ultimately for the Northern Ireland Ammunition Inspectorate, 321 EOD Unit, and the Weapons Intelligence Unit.

On paper this must have sounded fine. In practice it was not so simple. To start with, the word 'Intelligence' got in the way. A major was established at Headquarters Northern Ireland, a highly qualified officer who had previously commanded 321 EOD unit, but because the word appeared in his appointment he was absorbed into the normal 'Int' set-up run by the Colonel General Staff Intelligence (Colonel GS Int in military parlance). Now although he was an experienced soldier and had passed through Staff College, the Colonel GS Int was not 'weapons trained'. He was not a technician. As in all big concerns there is a tendency towards empire-building and HQ Northern Ireland was no different. With what I am sure were the best of intentions the Colonel GS Int actually suggested that CATO should come under his control as I was a gatherer of intelligence! We even had the nonsensical situation where the Colonel GS Int gave information about a device to the Commander Land Forces which was completely contrary to mine. The difference was that I had seen the device. He hadn't.

This did not make a lot of sense and things were not helped by the different ways in which the brigade HQs saw fit to use the captains sent to them as Weapons Intelligence Officers. To start with, not all the commanders were prepared to consider the captain's year with them as part of Staff training. This problem had to be dealt with by the Commander Royal Army Ordnance Corps in the Province – a lieutenant-colonel who was the senior representative of the Director General of Ordnance Services in the United Kingdom.

In 3 Brigade the plan originally suggested was actually implemented, and even improved upon. Possibly because of the pressures in South Armagh and Fermanagh, a genuine operational system was evolved and the Weapons Intelligence Officer, who was due to spend a year in the area as opposed to the four-month tour of the ATOs commanding the sections, virtually took over as EOD adviser to the brigade commander. This had a spin-off in taking some of the weight off the ATO's shoulders and freeing him for his main task of neutralising bombs. A Bomb Intelligence NCO was provided for all clearance operations mounted in 3 Brigade's area and he lived at Bessbrook on site with the RE Search Adviser and the ATO. This made a formidable team.

In Londonderry, 8 Brigade took a different view and used their captain as part of the overall intelligence set-up. They were not, however, particularly partisan about matters and did not seem to mind which way the argument went. As far as 39 Brigade were concerned, I got the impression that they didn't particularly want a Bomb Int officer.

To make matters even more complicated at that time, the Army in Northern Ireland was compelled to reduce its manpower requirements and the Royal Ulster Constabulary were restored to their normal duties as guardians of law and order with the Army's role to support them.

While the arguments continued about who should do the job of Weapons Intelligence the urgent requirement remained – to provide the research and development establishments in England, and other agencies, with details of the bomb capabilities of the IRA and other terrorist groups. Fortunately the major responsible for Bomb Intelligence at Headquarters Northern Ireland proved to be remarkably diplomatic and although his loyalty had to be to the Colonel GS Int (who would be responsible for writing his confidential report), he made sure that I saw all his reports, even though in some cases I had been left off the 'to see' distribution list. The reports were excellent and we were able to see dramatic changes in the types of explosive being used in various areas and pass back the information to England for the research and development establishments to produce new disruption devices and protective measures. The outcome was as vital for the protection of London as it was to the defence of Belfast.

Things were changing when I left Northern Ireland in August 1977, and both 8 Brigade and 39 Brigade were coming round to the system used in South Armagh. But much time had been wasted since a Weapons Intelligence Unit was first advocated, and at least one important research establishment confessed to me that it had been unaware that terrorist bombing tactics had changed so much over a period of three years.

Perhaps one way to have tackled the problem would have been to make sure that the Bomb Int officer at Headquarters Northern Ireland went on automatically to become CATO, thus providing the continuity that was so essential. It is usual for CATO to spend a year in the post and unless continuity exists his intimate involvement in the day-to-day bomb scene becomes his prime interest.

Having described differences which existed at Lisburn, I would not like anyone to think that they were due to anything other than a genuine desire on the part of everyone to beat the terrorist. One had to remember that the combat scene was changing and developing continually and differences of opinion were bound to occur. So were anomalies. In my own Corps the CRAOC was the senior representative in the Province of the Director General of Ordnanace Services and, as usual, boss of the Ordnance Depot. He was not, however, my boss as I reported to the Commander Land Forces. This could have posed problems in view of my seniority, but it didn't. When Maurice O'Dea came out as CRAOC on promotion to lieutenant-colonel we got on famously. When a problem arose about spares or cables for Wheelbarrow I asked him to let us have more entire equipments if spares were not available and he never failed us.

The Commander Royal Electrical and Mechanical Engineers (CREME) was another vital figure in our sphere. I had to deal with two during my tour. Andrew Maclaughlan and Chris Nitsch, and I could not have had better service. A blown-up Wheelbarrow would be rebuilt within about 12 hours and I recall that in the first three weeks of 1977 one section carried out 140 tasks of which one in seven involved Wheelbarrow.

REME fitters were provided in Lurgan and Londonderry and, when things got difficult, in Belfast as well. It was not unusual for the Belfast workshops to work 24 hours a day to get the indispensable Wheelbarrows back on the road. Some of the fitters were quite sentimental about the machines they serviced and if one was destroyed completely took it very hard indeed. They were quite aware of the importance operators attached to their equipment and as far as REME in Northern Ireland was concerned there was no such thing as an 'unauthorised' modification. After my experience with the petrol tankers two artificers worked non-stop for a weekend with me designing a rig to be used if similar circumstances should arise.

Among the civilians at Lisburn two figures stood out as far as I was concerned – one a squat Yorkshireman with bushy black eyebrows, Keith Norris, the Scientific Adviser, the other a tall, relaxed young man with a mop of hair, his deputy, Philip Haskell. On top of their other work, they gave invaluable advice on such things as the siting of bomb teams in preparation for the Orange Day Parades – it would have been pointless to have left them bottled up in Belfast. They could analyse, predict and calculate with infinite patience and good humour, and we were able to work together on providing various scenarios to the research establishments in England.

I cannot say that I looked on all senior civil servants at HQNI in the same friendly light, though with hindsight I appreciate the difficult position in which they found themselves from time to time. I was indiscreet enough to lend what, to me, was commonsense support to Lord Brookeborough when he tried to introduce a Private Member's Bill for the control and record of the millions of detonators produced every year by ICI. Unfortunately the government opposed the Bill – Lord Harris, then Minister of State at the Home Office considered that the Health and Public Safety at Work Act of 1974 was sufficient to deal with the detonator problem. I was officially ticked off for speaking publicly on a contentious political matter – as a serving soldier it was none of my business. The next time I was involved with politicians – I was invited to lunch with John Biggs-Davison and Airey Neave at the Commons in July 1977 – I got permission first.

I found myself less on the defensive when I became involved with the GOC's financial adviser, the Command Secretary, always referred to as 'Com Sec'. We had to do business together because of an incident about a year before I arrived in the Province when an NCO blew up some captured terrorist explosives in a disposal pit and caused damage to surrounding property. The farcical question was – who was going to pay for the damage? Was the NCO going to be mortgaged for life, or was the Ministry of Defence going to have to stump up? Was the damage really due to the professional irresponsibility of a government servant? Eventually it all revolved round one point: who was responsible for explosives when they were removed from their target by the EOD operator – the Northern Ireland Office or the Ministry of Defence? I made my position quite clear – if we were not granted immunity we would blow up all bombs *in situ* in future. Thus, if we had a case, which was not uncommon, where a bomb failed to explode because of some fault in its construction, we would achieve the terrorists' aim for them, at least as far as destruction of property was concerned. Someone somewhere saw the light and we were given immunity as long as we observed the code of practice of our trade when disposing of bulk explosives.

All claims by the public for compensation for damage caused by terrorist bombs were passed to me by the Northern Ireland Office at this time and either a sergeant or I dealt with them. We advised on the probability and type of damage likely to be experienced at various distances from the centre of explosion, taking into account the size and make-up of the bomb. The claims people reckoned we saved them many thousands of pounds, including one unusual claim from a farmer. A matter of years earlier a bomb had gone off a few miles from his hen houses and he was claiming that this had induced 'hysteria in his chickens' which had ceased to lay. Ah, well, it was a good try. . . .

173

CHAPTER FOURTEEN

Queen's Visit – King-Size Problems

I think it was Pat Lewis, the slightly irreverent Air Staff Officer Northern Ireland, who muttered darkly, 'If only the Devil could cast his net now. . . .'

We were standing on the roof of Coleraine University in august company: the GOC, Lieutenant-General Sir David House; the Commander Land Forces, Major-General Dick Trant;* the Chief Constable of Northern Ireland, the Deputy Chief, the Commander 8 Brigade, and a chief superintendent. On the floor beneath, out of sight, but just as interested in the search going on below, was the Secretary of State for Northern Ireland, Mr Roy Mason.

It was a balmy day with the temperature rising and the waters of the Bann sparkling beyond the trees. One could not have had better weather for a Royal occasion but . . . someone had reported that a bomb had been hidden somewhere and it was up to Mr Mason, in the final analysis, to decide whether or not it was right for the Queen to carry out her engagement to visit the university. Very properly, the Secretary of State had all the high-priced help he required on hand to give advice should he need it. Meanwhile the search teams patiently peered into tubs of brilliant flowers laid out to greet the Royal visitor. It was not a job for anyone who suffered from hay-fever.

The visit of the Queen and the Duke of Edinburgh to Northern Ireland for the celebration of her Silver Jubilee presented special problems for everyone.

The Royal party was not scheduled to go anywhere by road and the Royal yacht had to moor outside mortar or missile range of land. The Royal party flew in from their RN escort vessel by helicopter. However, free access by road had to be

* He had succeeded Major-General David Young.

provided for those travelling by car or coach to meet the Royal party and Hillsborough Castle and Coleraine University had to be searched and made secure. At Bangor and Portrush (which many holidaymakers know as the jumping-off point for the Giants' Causeway), where shipboard receptions were to be held, the departure points for the boats carrying guests to the yacht had to be searched above and below the water. Normal EOD cover had to be afforded the rest of the Province during the visit and finally any devices found during the Royal visit had to be cleared without alarming the Royal party. The last requirement could be construed only as, 'If you have to make a noise, make it quietly . . .,' and I decided this particular bridge would have to be crossed when we came to it. As for the other demands upon us, it was clear we would need reinforcements, extra ATOs who knew the area and the operational scene, plus a team of naval clearance divers.

I asked for and was promised without any quibbling three extra EOD teams and it did my heart good to see the names of those who volunteered and were selected, for they included two veterans who had served with me earlier, WO2 A— and Sergeant K—. My eyebrows did rise, however, on their arrival from England because they had been sent over equipped with Land Rovers and trailers which are excellent for use in an area where the population is friendly, but are not ideal in the context of Ulster. Furthermore their radios contained the wrong crystals and they could not operate on our frequencies.

We reorganised the radio situation and I solved the problem of the unsuitable vehicles by locating one Land Rover team within the protective cordon of troops which secured Hillsborough Castle and the second at Coleraine. The third was held in reserve at Lisburn.

The initial planning had begun weeks before the visit, much of it to the thunder of distant drums as the bands of Loyalist organisations prepared for the Orange marches of July and August. As the Lambegs boomed, I took the opportunity to make sure we had enough baton rounds in stock to deal with riot situations in July and any that developed during the Royal visit. I also had my car resprayed and the number plates changed. With the end of my tour in sight I saw no point in taking unnecessary risks.

As the period of the visit drew closer there was a marked

lack of intelligence information of particular bomb threats against the Royal party and it began to look as though we might not be bothered. Then, on 29 July, a device was found in a lavatory in Coleraine University, and neutralised. I was interested to note that it contained a new type of timer, first discovered in an arms cache in November. This worked electrically instead of mechanically, and its main advantage to the terrorist was that it could be set to function a considerable period after the bomb had been hidden and with a fair degree of accuracy regarding time. I included details of the timer in a talk to the International EOD Symposium but I fear I did not generate a great deal of interest in the possibility of having to counter what I called the Sleeping Bomb, first used in Canada.

The subsequent discovery of the Sleeper at Coleraine gave everyone food for thought because the grounds of the university were completely open, providing a major headache for the RUC and the Army authorities responsible. Searching of the surrounds began on 4 August and the police began scouring the buildings on the 7th.

I visited the university myself on the 9th to confirm arrangements, one EOD team being located outside the protective cordon and another inside, both coming under the operational command of the CO of the 2nd Coldstream Guards. As I left I told Chief Superintendent Nigel Spears jokingly that all would be well because I had visited the place. An hour after my departure a small bomb went off harmlessly in the grounds – another Sleeper with an electronic timer.

The following day, during the Queen's visit to Hillsborough, I decided to stay at HQ Northern Ireland a few miles away at Lisburn. I had complete confidence in the ATOs involved, and thought that, if anything, the opposition would attempt to cause problems on the main roads leading to the little town. They would have considered it a considerable propaganda victory if Her Majesty had flown in by helicopter and no one had been there to greet her. Two EOD teams were under operational command of the CO of the 1st Black Watch and I felt that if I was on the inside of the cordon and trouble flared outside I might have difficulty making a dignified exit from an area packed with sharp-eyed, suspicious Jocks. As it was, there was an alert because a 40-gallon oil drum was placed strategically on the Lisburn-Hillsborough road but this turned out to be a hoax (swiftly dealt with by a bomb-suited operator wielding a hook and line). The rest of the visit went smoothly though we were poised for action throughout. Fortunately nothing happened to mar the evening appointment I had been looking forward to.

For some time I had been in possession of an impressive piece of card which stated: 'The Master of the Royal Household has received Her Majesty's command to invite Lieutenant-Colonel D Patrick to a reception to be given by the Queen and the Duke of Edinburgh aboard HM Yacht Britannia off Bangor at 6.30 pm.' The receipt of this summons had caused some problems already, partly because it was clear that if terrorist activity flared up I would have to cry off, and partly because I was required to appear in Service Dress. When I went home for a weekend in July I tried on my SD for the first time in two years, and found out that I had actually put on weight. It didn't fit. Neither at the rear ... nor at the front. My wife set about frantically letting out the trousers and altering the tunic buttons. Next she experimented with bits of cardboard cut from a cornflakes box to make a stiffener for the silk ribbon of my one and only campaign medal which she then sewed on. I saw to it that the uniform was dry cleaned and thus attired for the first and only time during my Northern Ireland tour, and wearing a borrowed Sam Browne belt, I set off with two ATOs, a warrant officer and a sergeant, for Bangor.

In some ways the journey to the Britannia proved to be as exciting as a minor clearance ... certainly as interesting. After we had been searched by security men we guests were embarked on a flotilla of small boats, pinnaces from the yacht, RNR boats from Belfast, and locally hired motorboats. The further out we got, the more we realised there was distinct swell. By the time we reached the yacht, there was a rise and fall of two feet. Not all of the guests were young and fit and, while we queued to board, though they had our sympathy, the spectacle of them making the leap to the gangway at the side of the vessel provided as many laughs as a silent comedy. I was told later that no one took a bath but it must have been a near thing.

Aboard the Britannia, both the Queen and the Duke of Edinburgh talked to us about our job – the Queen as Colonel of the RAOC and Prince Phillip as a technical enthusiast, showing remarkable knowledge of our activities, equipment and our role. When they passed on to the next guests they left us feeling with pride that they knew they could rely on us to do our utmost to make their visit a success.

The next day, while we were awaiting the arrival of the Queen at Coleraine, the tip-off that there was another bomb around provoked instant reaction. Searching is a matter for experts and with the group of brasshats on the roof of the University, I watched keenly as the RE team and the dog handlers combed the area in which the Queen and the guests would be moving. In the meantime the arrival of the Royal party was delayed. Finally the reports were delivered to Mr Mason and, as nothing had been found, and the team using radio-controlled bomb detection equipment also drew a blank, the go-ahead was given. There was another minor scare when a cache of handguns was found on a road leading from Londonderry to Coleraine. The ubiquitous hook and line quickly sorted that one out as well. This visit also passed off perfectly.

As I have stated already, it was a fine summer's day and it seemed even lovelier once the Wessex helicopter carrying the Queen had lifted off safely and the happy crowds had disappeared.

'Ah, well, at least the difficult bit is over,' I said to Pat as we drove away. Five hours later a small bomb exploded harmlessly on some waste ground near the university – another Sleeper.

Leaving Northern Ireland, which I did for good soon after the Queen's visit, was a lot easier than I might have expected. The atmosphere was changing, familiar faces were vanishing from the Mess and the level of violence was gradually decreasing. Among my personal staff, Coupe was talking of what he would do in Civvy Street, and his co-driver, a tall, fair-haired Royal Corps of Transport lance-corporal called White, had gone off to Belize for six months. (He and Coupe used to compete for the dubious privilege of driving me to incidents.) Sniffer had been replaced, the OC of 321 EOD had moved on, my senior warrant officer was leaving, and, of course, the operators were always in the process of changing. Even the celebrated Flax Street Mill, for long the home of the Belfast EOD sections, had been closed and its inhabitants and their pin-up collection transferred to new quarters. True, the putting continued interminably on the green at Lisburn and the Harp Players were rehearsing for another production, but as the number of incidents fell and the RUC and the UDR took over more duties and responsibilities, a firmer grip was felt on the domestic affairs of the Mess at Lisburn. Among other things a strange black plastic box, nicknamed The Coffin, had been installed over the pump handles to prevent late-birds serving themselves with a drink after the bar had been closed. All in all, I suppose, things throughout the Province were improving. The farewell parties came and went and soon I was emptying the washing machine for the last time and packing my bags. Having handed over to my successor, and safely negotiated the ramp at the main vehicle checkpoint, I was on my way to Aldergrove to be flown home by courtesy of the Army Air Corps in a Beaver, known affectionately as TWA: 'Teeny Weeny Airlines'.

Returning to the mainstream of Army life proved to be much more difficult than quitting the Ulster scene. Because I had spent a dozen years in Germany, knew the people I would have to work for and how to please them and knew that the work would not be onerous, I had asked to go back there. Before I got there I had resigned my commission. After 25 years service I had decided that I had had enough and it was time to start another career in civilian life before I was too old. For my last three months of service I stayed at HQ DLSA.

Among my tasks was the completion of an ammunition storage study at HQ BAOR at Rheindahlen where I had to spend a few days. The first thing that struck me about the airport at Düsseldorf when I arrived was its air of opulence, reflected to some degree by the officers and soldiers passing through. The air of grubby tension one encountered at the docks in Belfast, the tight security, questioning looks and guarded conversations of Aldergrove, were missing along with the tired faces of soldiers and the weary, jumpy civilians. A smugness prevailed in BAOR. I was delighted to find Shirley Neild, promoted to lieutenant-colonel, installed there. I was also shaken to find that at that time she was about the only officer in 'A' Mess (for lieutenent-colonels and above) who had served in Northern Ireland. No one else seemed very interested in the place and I discovered if there was one way to clear the bar in the short time I spent there, it was to mention Ulster. There are, in fact, far more staff officers in BAOR who have not served in Northern Ireland than those who have, and the latter tend to talk among themselves.

There was far more interest in the great BAOR Jubilee Parade with all its ceremonial than in the brilliant success of the Queen in courageously visiting Northern Ireland.

Looking back I accept that in the eyes of most senior officers the campaign in Ulster was simply a nagging little ulcer, soaking up management and interfering with training for war in North-West Europe. A unit destined for a four-month tour of the Province has to undergo pre-op training and, after its tour, retrain as part of the brigade, battalion group, or combat team. Junior officers and men are frequently overworked and there have been problems with the families left behind. As far as my own Corps was concerned, the EOD commitment against the Irish terrorists was only a small part of the big picture, and they considered they had the right view on the priority to be afforded. I did not see things the way of my masters and resigned my commission to go into business (nothing to do with explosives), as have previous CATOs.

According to a respected long-term friend, a full colonel in the ammunition trade, CATOs returning from Northern Ireland were not themselves. He maintained we were irrational when we came home and as most commanders understood the problem our attempts to resign should have been ignored.

Perhaps he was right. Perhaps also the policy of learning as we go along, painful though it may be sometimes, has resulted in the British Army possessing not only soldiers trained for all-out war, but one of the best anti-terrorist organisations in the world. Specialists such as the RE search teams, the dogs, the EOD teams, and the SAS, combine dramatically to boost its effectiveness. Perhaps the problems posed by the bomber in Northern Ireland will be solved by training, selection, adequate supplies of the appropriate equipment, scientific support, and access to sound intelligence. My only plea to those responsible is that they should see that the man in the bomb suit is given only the best of everything. Nothing else is good enough.

APPENDIX

The Statistics of Terror – A summary of Bombing and Related Activity in Northern Ireland between July 1976 and July 1977

Month	Devices Exploded before EOD action	Devices Neutralised	Incendiary Devices Neutralised	Hoaxes	False Alarms	Finds of Explosive
1976						
Jul	80	32	13	28	59	51
Aug	78	40	33	38	84	52
Sep	56	36	17	132	64	35
Oct	82	26	27	30	84	37
Nov	33	18	8	22	90	35
Dec	45	28	36	44	90	26
1977						
Jan	53	29	54	32	92	38
Feb	23	18	28	29	71	40
Mar	50	24	14	39	87	56
Apr	53	19	10	25	80	64
May	33	15	38	146	63	56
Jun	17	8	5	7	44	37
Jul	7	7	10	27	44	33

The EOD teams were dealing with an average of more than three bombs a day during this period.

Index